SCHOLAST

GRADE
6

PROBLEM SOLVED:
BAR MODEL MATH

Bob Krech

New York • Toronto • London • Auckland • Sydney
Mexico City • New Delhi • Hong Kong • Buenos Aires

Editor: Maria L. Chang
Cover design by Tannaz Fassihi
Cover art by Matt Rousell
Interior design by Grafica Inc.
Interior illustrations by Mike Moran

ISBN: 978-0-545-84014-9
Copyright © 2016 by Scholastic Inc.
All rights reserved.
Printed in the U.S.A.
First printing, June 2016.

1 2 3 4 5 6 7 8 9 10   40   22 21 20 19 18 17 16

Table of Contents

Introduction

The very first process standard outlined by both the National Council of Teachers of Mathematics (NCTM) and the more recent Common Core State Standards (CCSS) focuses on understanding problems and "persevering in solving them." They purposefully listed this standard first because problem solving is what teaching math is all about. All of the skills, concepts, knowledge, and strategies students learn in math are basically tools. The eventual goal is that students will apply these math tools to solve problems they encounter in life.

Problem Solved: Bar Model Math is designed to help you and your students learn about a new problem-solving tool—a versatile and effective strategy commonly known as Bar Modeling. A major component of Singapore Math, Bar Modeling has been proven effective in helping students achieve high levels of mathematical competency.

For more than two decades, Singapore's students have consistently ranked among the highest in the world in international math assessments, such as TIMSS (Trends in International Mathematics and Science Study) and PISA (Programme for International Student Assessment). Looking at this success, many schools and districts in the United States and around the world have begun to examine and use ideas found in the Singapore Math curriculum. In our opinion, Bar Modeling is one of the most powerful—if not *the* most powerful—component of Singapore Math.

What Is Bar Modeling?

Bar Modeling is a unique and incredibly versatile strategy that can be used effectively by the youngest elementary school children all the way through to college math majors. This strategy can be applied to a wide range of problem types and contexts. We typically teach students many different strategies (such as Look for a Pattern, Draw a Picture, or Guess and Check) for tackling word problems. With the Bar Modeling method, we can attack any of the various problem types with one singular, powerful approach.

In Bar Modeling, we break out the information in a problem and represent it in a simplified pictorial form using bars or rectangles to show quantities. The pictorial representation helps students better see and understand the quantities in—and thus the possible solutions to—the problem. The great advantage of the method is that students can visually represent both the given facts in the problem as well as the "unknown" (what they are trying to find out) in a way that allows them to *see* relationships between the quantities, thus promoting flexible thinking about numbers and general number sense. Bar Modeling empowers students to become thinkers, not memorizers.

Problem-Solving Methodology

Two key parts of any problem-solving methodology include identifying the relevant information and determining the question that needs to be answered. Throughout this book, students will be asked to underline relevant information and circle the question. However, if your class already uses a different problem-solving method, continue to use the procedure they've been taught and modify the directions of these lessons accordingly.

Problem Solving and Bar Modeling?

There are many ways to describe a consistent, logical problem-solving approach, but most experts would agree it includes some variation of the following:

1. Carefully read directions to determine the real-world context of the information in the problem and what the problem is asking to be solved.
2. Determine which data are relevant and which are not.
3. Select a strategy to solve the problem.
4. Translate the English language directions into mathematical form and accurately solve the problem.
5. Review the result for reasonableness and accuracy.

It is in the "translate English into math" part where bar models can be particularly helpful. In 6th grade, students are making the transition from arithmetic in the lower grades to algebra in the upper grades. Many students have difficulty taking the real-world scenario in a word problem and depicting it directly in equation form. Setting up a simplified representation of the problem helps support students, particularly visual learners, as they move from reading and interpreting directions to solving the problem.

In addition to problem solving, other practice standards expect students to "model with mathematics" and "use appropriate tools strategically." The Bar Modeling approach helps students meet these expectations. Furthermore, in 6th grade, CCSS content standard 6.RP.A.3 specifically notes that students should be able to reason about "tape diagrams" (i.e., bar models) as they solve ratio and proportion problems. Teaching Bar Modeling is an effective way to meet a number of content and practice standards.

How Does Bar Modeling Work?

When first learning how to solve word problems, young students find it very helpful to have problems represented by physical manipulatives, such as cubes or blocks. Later, students might draw pictures or even squares to make the representation a little more abstract.

Bar Modeling takes the pictorial representation a step further toward the abstract. Students represent quantities by drawing simple rectangles or bars labeled with words, arrows, and numbers. This transition from drawing actual objects to simpler representations is particularly important since quantities eventually become so large it is impossible to draw or represent them pictorially. A simple rectangle or bar with appropriate labels can do the job.

At the same time, a number of middle school students aren't quite ready to depict and solve problems algebraically. Bar Modeling can help students make the transition from concrete to abstract by giving them a method for representing quantities and situations in a way that is more abstract than pictures or counters, but also more concrete than equations.

Bar Modeling Variations

Bar Modeling can be applied to a wide range of math problems. There are variations that suit different kinds of problems, but the Part-Whole model and the Comparison model are the two basic types of bar models. Be aware, however, that more complex problems may require using a series of models to solve the problem.

Part-Whole Model

Consider this problem: *Farmer Mary has 122.3 acres allocated to growing organic apples. She decides to convert 9.7 acres to grow organic salad greens for local farm-to-table restaurants. How many acres are now used to grow organic apples?*

We have two parts in this problem: the number of acres converted to growing salad greens (9.7 acres) and the number of acres now allocated to growing apples (the unknown in the problem). The "whole" in this problem is the number of acres on the farm (122.3 acres).

To represent this problem, begin by drawing a bar to represent the whole, labeling it with the quantity and drawing an arrow to show that the quantity applies to the entire bar:

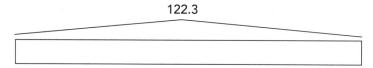

Next, divide the bar into two parts to represent the number of acres growing apples (unknown) and salad greens (9.7):

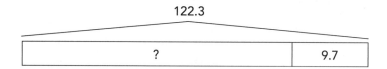

The parts add up to a whole (? + 9.7 = 122.3), and the missing part can be found by subtracting the known part from the known whole: 122.3 − 9.7 = 112.6.

The Part-Whole model can also be used for addition, multiplication, and division problems where there are parts and a whole, and one of the pieces is unknown.

Comparison Model

Consider this other problem: *Farmer Mary's farm has a Brussels sprouts yield of 128.5 bushels per acre, while Farmer David's farm has a yield of 112.6 bushels per acre for the same crop. How much greater is Farmer Mary's yield?*

Here's how to represent this problem using a Comparison Model: First, draw and label two bars to show the yields for the two farms. Since Mary's yield is greater than David's, her bar should be longer.

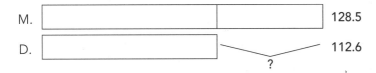

| M. | | 128.5 |
| D. | | 112.6 |

Next, use an arrow to show that the difference in length between the two bars represents how much greater Mary's yield is than David's, i.e., the difference between the two yields. Show the unknown with a question mark.

| M. | | 128.5 |
| D. | | 112.6 |

Finally, turn the representation into a calculation. It's apparent from the diagram that the unknown is the difference between the two bars: 128.5 – 112.6 = 15.9 bushels per acre.

What Is in This Book?

This book offers a step-by-step approach to helping students master Bar Modeling techniques through solving problems appropriate for 6th grade students. Accordingly, topics like fractions/decimals/percentages, ratio and proportion, and algebra are featured. To help students learn the basics of the technique, addition/subtraction and multiplication/division problems— operations students have more experience with—are used in the first two lessons.

There are 20 lessons in all, each with four carefully chosen word problems that gradually increase in difficulty. Use the first two problems in each lesson to introduce the strategy, then have students practice it on their own with the other two problems. All of the word problems are presented in two ways:

- A digital version can be accessed through **www.scholastic.com/ problemsolvedgr6**. (Registration is required.) Display the problems on the interactive whiteboard during your lesson to support your teaching.
- A reproducible version for students can be found in the second part of this book, starting on page 56. These pages feature a lightly printed graph-paper background to help students keep their diagrams neat and organized.

As you work with students on the various word problems, follow the problem-solving process outlined at the top of page 5. Repeating this routine regularly will help students confidently tackle any problem they encounter, especially in testing situations. By the time you have completed the lessons in this book, you will have equipped your students with another powerful tool they can count on as they continue to develop their abilities as excellent, world-class problem solvers.

A Middle School Sense of Humor

As you are no doubt aware, 6th grade students have a "certain" sense of humor. Studies indicate that student engagement increases when humor is evident. Accordingly, you will not find problems about two trains that each leave Kansas City at such and such a time or a swimming pool that is being filled at such and such a rate. The problems in this book were written so students will enjoy and look forward to solving them.

Bar Modeling With Basic Operations

This chapter introduces students to the concept, conventions, and vocabulary of Bar Modeling. By 6th grade, most students have been adding and subtracting multi-digit numbers and should be comfortable with these operations using the conventional algorithms. By using these familiar operations in the first two lessons, students won't have to engage with new content and can focus their attention on learning how to use Part-Whole and Comparison bar models.

LESSON 1

Addition and Subtraction Problems

Materials: student pages 56–57, pencils, projector, interactive whiteboard, markers

Preparation: Distribute copies of pages 56–57 and pencils to students. Go to www.scholastic.com/problemsolvedgr6 and click on Lesson 1. Set up your computer and projector to display the problems on the interactive whiteboard.

As your class works on the problems throughout this book, let students know that they are expected to draw bar models and write matching equations to show their work for solving each problem.

Display Problem #1 on the interactive whiteboard.

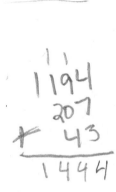

Prunella was admiring her dental floss collection. She has 207 pieces of mint-flavored floss, 1,194 pieces of cinnamon-flavored floss, and 43 pieces of extremely rare kumquat-flavored floss. How many pieces of dental floss does Prunella have in all?

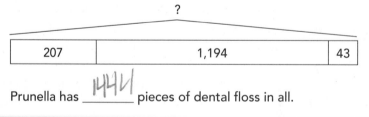

| ? |
| 207 | 1,194 | 43 |

Prunella has _____ pieces of dental floss in all.

Read aloud the problem. Ask students to identify and underline the relevant facts (the three addends) on their papers. Then have them identify and circle the question as well (the number of pieces of dental floss in all).

Ask the class how to solve the problem. *(Set up a traditional three-addend addition problem: 207 + 1,194 + 43.)* Then have them solve the problem. *(Find the sum, which is 1,444.)* Call on a volunteer to write the equation on the board.

Tell students that you will show them a different technique they can use to help them solve problems. This technique is called Bar Modeling, and it uses bars to *represent* quantities and relationships between quantities. Tell the class you are going to show them how to solve the problem using a Part-Whole model.

Point out the relevant information the class underlined in the problem, i.e., 207 pieces of mint floss, 1,194 pieces of cinnamon floss, and 43 pieces of kumquat floss. First, draw and label a bar for the 207 mint-flavored pieces on the whiteboard. Note that the bar *represents* the quantity of 207 pieces of mint floss. (See below.)

Then, one at a time, show how to add each of the other parts to the initial bar to arrive at the sum. Add the cinnamon-flavored floss . . .

Then the kumquat-flavored floss . . .

Finally, add an arrow and question mark to indicate that we are looking for the total of the three parts, as shown below.

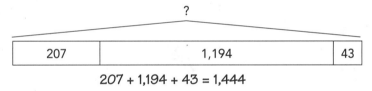

$$207 + 1,194 + 43 = 1,444$$

Point to the traditional solution on the board from earlier in the lesson (207 + 1,194 + 43 = 1,444) and show how each part of the model represents the parts of the problem and the solution.

Explain that another option is to draw the addends one above the other, as shown below.

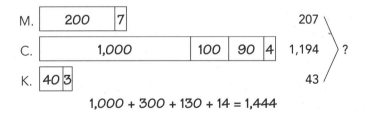

$$1,000 + 300 + 130 + 14 = 1,444$$

With this second option, you may also want to demonstrate to students how they can section each bar into place values (i.e., 200 + 7 for the first bar; 1,000 + 100 + 90 + 4 for the second bar; and 40 + 3 for the third bar) and then add the numbers in each place and combine these partial sums to get to the total.

Bars Are Approximate

When students begin to draw bars, they will probably use the graph paper squares to make exactly numbered bars. Help them understand that a bar that is 2 inches long, for example, could represent the quantity 100. Students should try to *approximate* the size of the bars in the same problem in relation to each other. For example, if a problem has the amounts 78 and 32, ask students: *Is 32 about half of 78? A quarter of 78?* This is actually good number sense and mental math practice. Help students understand that the 32 bar should be a little less than half of the 78 bar.

Explain to students that while they might find it quicker to solve this problem using the standard algorithm and without having to draw bars, the bar model technique will come in handy later when they are assigned more complex problems. Tell the class that to learn how Bar Modeling works, we are using problems we already can solve using traditional methods.

Display Problem #2 on the interactive whiteboard.

It's the first day of business for Francine's Fez shop. Prior to the big day, Francine worked for 5 days in a row, 18 hours each day, making high-quality fezzes. By the time the doors opened, she had made 947 fezzes. After serving a happy throng of customers, she had only 159 fezzes left. How many fezzes did she sell on her first day of business?

947	
159	?

Francine sold _____ fezzes on her first day of business.

Read aloud the problem. Ask students to identify and underline the relevant facts (the number of fezzes at the beginning and at the end of the day) on their papers. You might also want to have students identify data in the problem that is <u>not relevant</u> (the number of days and hours per day worked) and cross this out. Finally, ask students to identify and circle the question the problem wants answered (the number of fezzes sold on the first day).

Guide students to understand that in this problem we are given a total (the whole) and one of the addends that make up the whole. The other addend is unknown. Ask the class for suggestions on how to solve the problem using a traditional approach. *(Set up the equation 947 – x = 159 and manipulate the equation to 947 – 159 = 788.)* Call on a volunteer to write the equation on the board. If necessary, show students how 947 – *x* = 159 can be manipulated to 947 – 159 = *x*.

Then challenge students to work individually or in pairs to represent and solve the problem using a Part-Whole bar model. Ask volunteers to share their work on the board. Models should look something like the diagram above.

Show how the parts of the model correspond to the traditional equation 947 – 159 = *x*, with *x* = 788. Explain that if 947 is the whole and 159 is part of the whole, then the difference between 947 and 159 must be the missing part. Point to the traditional subtraction equation written on the board earlier in the lesson and show how the parts of the equation have a corresponding part in the bar model. Note that the problem is still solved by subtracting 159 from 947, but that setting up the bar first makes it easier to take the wording from the problem and set up a calculation.

This diagram format also makes it easy to add up to find the difference between the two numbers. If we think about it as a number line, we can jump up 1 from 159 to 160. Go up 40 more, and we're at 200. 700 more brings us to 900, then add on 47, and we're at 947. When we add these jumps, we get 1 + 40 + 700 + 47 = 788, which is the difference.

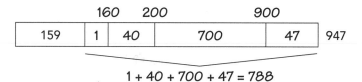

Have students work on Problems #3 and 4 (page 57) independently or in pairs. Tell students they will need to draw bars to represent the amounts in each problem. Give students a few minutes to work. Then display each problem on the whiteboard. Call on volunteers to draw the diagrams and write equations on the board. Then invite students to share their strategies for solving the problems, interacting on the whiteboard whenever possible.

LESSON 2

Multiplication and Division Problems

Materials: student pages 58–59, pencils, projector, interactive whiteboard, markers

Preparation: Distribute copies of pages 58–59 and pencils to students. Go to www.scholastic.com/problemsolvedgr6 and click on Lesson 2. Set up your computer and projector to display the problems on the interactive whiteboard.

As students work with bar models and multiplication, remind them that repeated addition is the basis of multiplication, which will be apparent in the diagrams they draw.

Students might find the following summary useful when using bar models to solve multiplication and division problems.

Part-Whole Model

One part × the number of parts = the whole
The whole ÷ the number of parts = one part
The whole ÷ one part = the number of parts

Comparison Model

The larger bar ÷ the smaller bar = the multiple
The smaller bar × the multiple = the larger bar
The larger bar ÷ the multiple = the smaller bar

Irrelevant Information

Students who have been taught an algorithm but don't really understand the underlying concept will look for numbers from the problem to plug into the algorithm without thinking about whether they make sense or not. That's why it's important to promote the good habit of underlining relevant information in a word problem and crossing out irrelevant information.

Display Problem #5 on the interactive whiteboard.

Celebrity Tina Trashtruckian's living room has 27 mirrors so she can always see how she looks from any angle. The room also has 3 times as many pictures of the person Tina loves most in this world—herself—as it does mirrors. How many pictures of Tina are in the living room?

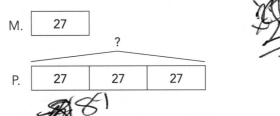

M. | 27 |

?

P. | 27 | 27 | 27 |

There are ~~72~~ *81* pictures of Tina in the living room.

Read aloud the problem. Have students underline the relevant information (27 mirrors, 3 times as many pictures as mirrors) and circle the question (how many pictures in all) on their papers.

Ask the class if they think a Part-Whole bar model would be appropriate for this problem. *(No)* Point out that this problem doesn't give us a "whole" and a "part." Explain that there is another type of bar model called a Comparison model, which, as its name implies, allows us to compare two or more quantities. Ask: *What are we comparing in this problem?* (Mirrors and pictures) Explain that in this problem one of the quantities is a multiple of the other. Demonstrate on the board how the information in the problem can be represented in bar format, as shown above.

Our unit value is 27. The number of mirrors has one unit of 27 while the number of pictures, has 3 equal units of 27. To find the total number of pictures, multiply 27 (the unit) by 3 (the number of units): 3 × 27 = 81 pictures.

We could use partial products again, splitting each unit in the bar into 20 and 7, so 3 × 20 = 60 and 3 × 7 = 21. Add those partial products together: 60 + 21 = 81.

Display Problem #6 on the interactive whiteboard.

Estelle's secret recipe for caramel beet dainties calls for 39 beets. For the coming holidays, she's going to make a batch and split it among 3 friends. How many beets will be in the gift each friend receives?

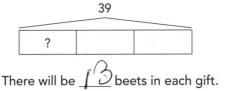

39

| ? | | |

There will be ___ *13* beets in each gift.

Read aloud the problem, then ask students to underline the relevant information (39 beets, 3 friends) and circle the question (how many beets will each friend receive) on their papers. Ask the class: *Which is more appropriate for this problem—a Part-Whole bar model or a Comparison bar model?* (Part-Whole, since we know the total and the number of parts, but don't know the amount of each equal part)

Challenge students to work individually or in pairs to set up and solve this problem. Then call on volunteers to share their diagrams and solution strategies on the board. As needed, demonstrate how to solve the problem using a Part-Whole model, as shown below.

$$39 \div 3 = 13$$

Explain that since we have 3 equal groups of an unknown amount and 39 beets in all, we can find the number in each group by dividing the total number of beets (39) by the number of groups (3): $39 \div 3 = 13$ beets in each group.

Have students work on Problems #7 and 8 (page 59) independently or in pairs. Remind them to draw bars and write equations to represent each problem. Give students a few minutes to work. Then display each problem on the whiteboard. Call on volunteers to draw the diagrams and write equations on the board. Then invite students to share their strategies for solving the problems, interacting on the whiteboard whenever possible.

Bar Modeling With Fractions, Decimals, and Percentages

This chapter introduces the concept of using bar models to represent and solve fraction, decimal, and percentage problems. The previous lesson on multiplication and division will inform this work as fractions are an expression of division, and students will readily see this as they diagram and solve these problems.

LESSON 3

Fraction Basics

Materials: student pages 60–61, pencils, projector, interactive whiteboard, markers

Preparation: Distribute copies of pages 60–61 and pencils to students. Go to www.scholastic.com/problemsolvedgr6 and click on Lesson 3. Set up your computer and projector to display the problems on the interactive whiteboard.

The topic of fractions can be difficult for some students, in part because of the many ways fractions can be viewed. A fraction can be seen as a division problem, a part of a whole, a part of a number (1/3 of 12 = 4), or a part of a set. Fraction notation is also one way to express ratios; for example, 1:3 can also be written as 1/3. Remind students that when reading the directions to a problem with fractions, they should determine the meaning of the fraction in the context of that particular problem.

Display Problem #9 on the interactive whiteboard.

Clementine had 18 pet wolverines. Her mother insisted that she give them away. Clementine had no choice but to obey and donate them to nature centers. If each nature center will receive 1/6 of Clementine's pack, how many wolverines will go to each center?

18

?					

Each center will receive __3__ wolverines.

Read aloud the problem. On their papers, have students circle the question (how many wolverines will each nature center receive) and underline the relevant information (18 wolverines in all, each center receives 1/6 of the wolverines). Ask the class: *Would it be better to approach the problem with a Part-Whole bar model or a Comparison bar model?* (Part-Whole, since we are given the whole and asked to find a part)

Demonstrate on the board how to tackle this problem. Start with a bar to represent the whole (18). Since Clementine is giving each center 1/6 of her pack, divide the bar into 6 sections, as shown on page 14.

If 6 sections equal 18, then one section equals 18 ÷ 6 = 3. Students can see from the diagrams how division and fractions are interrelated and how retrieving a multiplication fact helps immensely with fraction work because of this relationship.

18

3	3	3	3	3	3
1/6	1/6	1/6	1/6	1/6	1/6

Display Problem #10 on the interactive whiteboard.

Petunia had $280. To make a powerful fashion statement, she spent 5/7 of her money on a pioneer-style sunbonnet, created by celebrity designer Dieter. After paying for the sunbonnet, how much money did Petunia have left?

280

S	S	S	S	S	40	40

?

Petunia had $ _80_ left.

Fractions Chart

As students work with fractions, enlist their help in creating a fractions reference chart. Start with important benchmark fractions, such as 1/4, 1/2, 1/10, and 3/4. Write each as a numeral, in word form, and with a small diagram. Then have students add other fractions they encounter in problems and on packages and advertisements to build a real-world connection.

Read aloud the problem. On their papers, have students circle the question (how much money Petunia had left after buying the sunbonnet) and underline the relevant information ($280 to start, 5/7 of her money spent on the sunbonnet). Ask the class: *Would it be better to approach the problem with a Part-Whole model or a Comparison model?* (Part-Whole, since we are given the whole and asked to find a part)

Challenge students to work individually or in pairs to represent the problem using a bar model and solve it. Then call on volunteers to share their diagrams and strategies for solving the problem on the board. The solutions should follow this reasoning:

Start with a bar to represent the whole, $280. Next, divide the whole into 7 units because the problem is talking in terms of sevenths. Five of these units will represent the cost of the sunbonnet. (See above.) If there are 7 sections and the total equals $280, then each section equals $280 ÷ 7 = $40. So the money left after the purchase of the sunbonnet equals $40 + $40 = $80.

280

S	S	S	S	S	$40	$40

$280 ÷ 7 = $40   2 x $40 = $80 $80

Have students work on Problems #11 and 12 (page 61) independently or in pairs. Remind them to draw bars and write equations to represent each problem. Display each problem on the whiteboard, then call on volunteers to share their work and strategies on the board.

LESSON 4

Using Part-Whole Models to Solve Fraction Problems

Materials: student pages 62–63, pencils, projector, interactive whiteboard, markers

Preparation: Distribute copies of pages 62–63 and pencils to students. Go to www.scholastic.com/problemsolvedgr6 and click on Lesson 4. Set up your computer and projector to display the problems on the interactive whiteboard.

Since by definition a fraction is part of a whole, it's only logical that most fraction problems can be represented by Part-Whole bar models, as students will see in this lesson.

 Display Problem #13 on the interactive whiteboard.

Manfred had his heart set on buying the bestseller, *Everything I Needed to Know, I Learned From My Gerbil.* He spent 3/10 of his savings on the book, which cost $21.21. How much did he have in savings before he made the purchase?

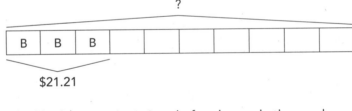

Manfred had $_____ in savings before he made the purchase.

Read aloud the problem, then have students circle the question (Manfred's savings before he purchased the book) and underline the relevant information (the book cost $21.21, which made up 3/10 of his savings) on their papers. Ask the class: *Does this problem call for a Part-Whole bar model or a Comparison bar model?* (Part-Whole, since we're given parts and are asked to find the whole)

Demonstrate on the board how to go about solving this problem. Start by setting up a bar to represent the whole, i.e., the amount of Manfred's savings before he bought the book. Then ask: *Into how many parts should the whole be divided?* (10, since we know that 3/10 of Manfred's savings was used to buy the book) Since the $21.21 cost of the book amounts to 3/10 of his savings, we can represent that as follows:

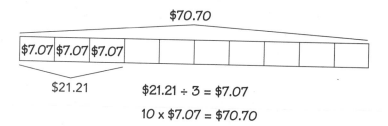

$$\$21.21 \div 3 = \$7.07$$
$$10 \times \$7.07 = \$70.70$$

Since 3/10 of Manfred's savings equals $21.21, we know that 1/10 of the savings equals $21.21 ÷ 3 = $7.07. Because the bar has been divided into 10 sections and each one equals $7.07, we know that Manfred's total savings equals 10 × $7.07 = $70.70.

Display Problem #14 on the interactive whiteboard.

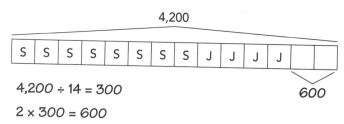

4,200 people auditioned for the new hit TV show, *I'm Dying to Be on Television!* 2/7 of the people in line were clumsy jugglers, 8/14 of the contestants were off-key singers, and the rest were humorless comedians. How many were humorless comedians?

4,200

| S | S | S | S | S | S | S | S | J | J | J | J | | |

?

There were _____ humorless comedians.

Read aloud the problem. On their papers, have students circle the question (the number of humorless comedians) and underline the relevant information (4,200 people were in line, 2/7 of them were clumsy jugglers, 8/14 of them were off-key singers, the rest were humorless comedians). Ask the class: *Is a Part-Whole model or a Comparison model more appropriate for this problem?* (Part-Whole, because we know the whole and are looking for a part)

Challenge students to work individually or in pairs to represent the problem with a bar model and solve it. Call on volunteers to share their diagrams and strategies for solving the problem on the board.

One way to begin is to draw a bar representing the 4,200 people in line. Then focus on the parts of the whole. We know that 2/7 are jugglers and 8/14 are singers. Point out that 14 is a multiple of 7 and that 2/7 can be converted to the equivalent fraction 4/14. Since all the parts now have a common denominator, we can divide the bar into 14 parts, labeling 8 parts as singers and 4 parts as jugglers.

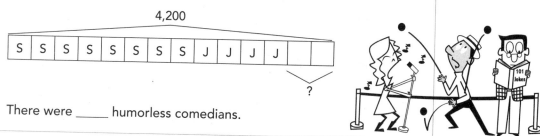

4,200

| S | S | S | S | S | S | S | S | J | J | J | J | | |

$$4,200 \div 14 = 300$$
600
$$2 \times 300 = 600$$

When teaching a multiplication fact such as 3 × 2, emphasize to students that 3 × 2 means "three groups of two." Explain that the "×" symbol translates to "groups of."

When working with fractions, use the same phrasing. If we need to find 1/2 of 36, for example, essentially we are doing 1/2 × 36 or "half a group of thirty-six," which is 18. This type of thinking and wording will help students as they find fractions of numbers.

Note that if the whole equals 4,200 and there are 14 parts, then each part equals 4,200 ÷ 14 = 300 people. Since humorless comedians make up 2/14 of the whole, we can find their number simply by multiplying 2 × 300 = 600.

Have students work on Problems #15 and 16 (page 63) independently or in pairs. Remind them to draw bars and write equations to represent each problem. Display each problem on the whiteboard, then call on volunteers to share their work and strategies on the board.

LESSON 5

Using Comparison Models to Solve Fraction Problems

Materials: student pages 64–65, pencils, projector, interactive whiteboard, markers

Preparation: Distribute copies of pages 64–65 and pencils to students. Go to www.scholastic.com/problemsolvedgr6 and click on Lesson 5. Set up your computer and projector to display the problems on the interactive whiteboard.

Indicate to the class that while Part-Whole bar models are generally useful for solving problems involving fractions, they aren't always the best way to solve a problem. When two or more quantities are being compared, a Comparison bar model is obviously preferable, as they'll see in the following problems.

Display Problem #17 on the interactive whiteboard.

Gilbert's father ordered him to dispose of his overly ripe, smelly cheese collection. He gave 2/5 of his collection to the Hold-Your-Nose Cheese Emporium and 1/3 of the collection to the school cafeteria. If the Cheese Emporium received 30 more pieces of cheese than the cafeteria, how many pieces of cheese did the cafeteria get?

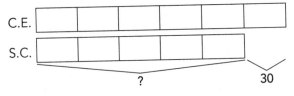

The cafeteria got _____ pieces of cheese.

Read aloud the problem. On their papers, have students circle the question (how many pieces of cheese did the cafeteria get) and underline the relevant information (Gilbert gave 2/5 of his cheese collection to the Hold-Your-Nose Cheese Emporium and 1/3 to the school cafeteria; the Cheese Emporium received 30 more pieces than the cafeteria).

Start by pointing out that it would be helpful to find a common denominator for the two fractions in the problem. Ask: *What is the least common denominator?* (15) Since 15ths is the least common denominator, we can convert the Cheese Emporium's 2/5 of the collection to 6/15 and the cafeteria's 1/3 to 5/15. Because our unit is a 15th of Gilbert's collection, we can set up our Comparison bar model as shown on page 18, with the Cheese Emporium getting 6 units and the school cafeteria getting 5 units.

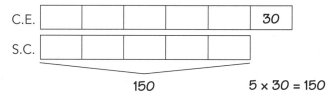

We know that the Cheese Emporium received 30 more pieces of cheese than the cafeteria and has one more unit than the cafeteria. That means the value of one unit is 30 pieces of cheese. Since the cafeteria received 5 units, it received 5 × 30 = 150 pieces of cheese.

Display Problem #18 on the interactive whiteboard.

> Wilhelm the Worm was comparing his length to that of his siblings. Wolfric is 1/3 of Wilhelm's length, and Waldo is twice as long as Wilhelm. If Wolfric is 12 centimeters long, how long is Waldo?
>
> Wolf. | 12
> Wilh.
> Waldo
> ?
>
> Waldo is _____ centimeters long.

Read aloud the problem. Have students circle the question and underline the relevant information on their papers. Ask: *What kind of bar model does this problem call for?* (Comparison, since we are comparing three different lengths) Challenge students to work individually or in pairs to represent the problem with a bar model and solve it. Call on volunteers to share their diagrams and solutions on the board.

Here is one way to approach this problem: First, set up bars to compare the three worms and record Wolfric's length of 12 centimeters, as shown above.

Check that the bars have the correct number of units as specified by the problem: Wolfric has 1/3 of Wilhelm's units, and Waldo has twice as many units as Wilhelm. Since Wolfric's one unit has a value of 12 cm and Waldo has a length of 6 units, that means Waldo is 72 cm long (6 × 12).

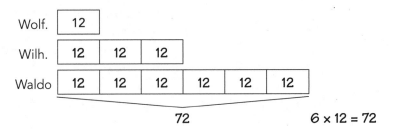

$$6 \times 12 = 72$$

Homework

If you want students to practice these strategies at home, assign developmentally appropriate word problems from any source. Ask students to solve the problems using the Bar Modeling method that you have been practicing in class. Provide graph paper or horizontally lined paper to help students with their drawings.

Have students work on Problems #19 and 20 (page 65) independently or in pairs. Remind them to draw bars and write equations to represent each problem. Display each problem on the whiteboard, then call on volunteers to share their work and strategies on the board.

`LESSON 6`

Dividing Fractions

Materials: student pages 66–67, pencils, projector, interactive whiteboard, markers

Preparation: Distribute copies of pages 66–67 and pencils to students. Go to www.scholastic.com/problemsolvedgr6 and click on Lesson 6. Set up your computer and projector to display the problems on the interactive whiteboard.

The concept of what it means to divide by a fraction is difficult for many students to grasp. Students often rely on the memorized instruction to "invert and multiply" without really understanding what it is they are doing. Bar models can be particularly useful here because they help students visualize the real-world situation depicted in the problem.

Present students with a relatively simple problem with a whole number divided by a fraction. For example: *A carpenter has a 2-foot long board and wants to cut it into 1/4-foot pieces. How many 1/4-foot pieces can he cut from the board?* Draw a board, labeling the width as 2 feet, then show how the board can be cut into 8 different 1/4-foot pieces by drawing vertical lines, as shown below.

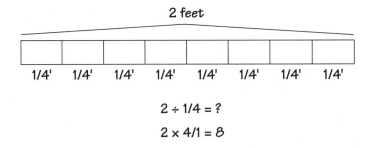

$$2 \div 1/4 = ?$$
$$2 \times 4/1 = 8$$

After the class understands the visual representation of a 2-foot board cut into 8 1/4-foot pieces, review the algorithm, i.e., "invert and multiply": $2 \div 1/4$ is the same as $2 \times 4/1 = 8$.

Provide another example, this time with a mixed number divided by a fraction: *The same carpenter has a 1-1/2-foot board and wants to cut it into 1/4-foot pieces. How many pieces will he cut from the board?* Repeat the process as before, this time showing that the result will be 6 different 1/4-foot pieces.

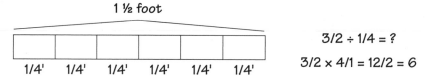

1 ½ foot

1/4' 1/4' 1/4' 1/4' 1/4' 1/4'

$3/2 \div 1/4 = ?$

$3/2 \times 4/1 = 12/2 = 6$

After the class understands the visual representation, review the algorithm.

Display Problem #21 on the interactive whiteboard.

Celebrity Dustin Dweeber ordered 2 pizzas from his favorite pizza shop for his posse. 1-1/2 of the pies will have a special gold-leaf topping and 1/2 pie will have plain cheese topping for Angus, Dustin's bodyguard, who doesn't eat precious metals. Dustin asked the shop to cut each pie into sixths. How many slices of gold-leaf pizza will the pizza shop make?

P.1 [   ]   P.2 [| C]

The pizza shop will make _____ slices of gold-leaf pizza.

Read aloud the problem, then have students circle the question and underline the relevant information on their papers.

Start by drawing two bars to represent the two pizzas. Label half of one of the bars with C for cheese, representing Angus's portion. (See above.) Then draw vertical lines in the two bars to divide each pie into 6 portions. Label all 6 portions on the first pie with GL for gold leaf and 3 portions with GL and 3 with C for cheese on the second bar.

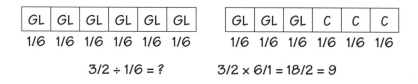

| GL | GL | GL | GL | GL | GL |
| 1/6 | 1/6 | 1/6 | 1/6 | 1/6 | 1/6 |

| GL | GL | GL | C | C | C |
| 1/6 | 1/6 | 1/6 | 1/6 | 1/6 | 1/6 |

$3/2 \div 1/6 = ?$ $3/2 \times 6/1 = 18/2 = 9$

We can see that there are 9 slices with gold-leaf topping. Show how this would also work with the standard "invert and multiply" approach: $3/2 \div 1/6 = 3/2 \times 6/1 = 9$.

Display Problem #22 on the interactive whiteboard.

Sports nutritionist Dr. Ferdie Fumble has developed a new bologna-flavored sports drink. His first test batch is 3/4 of a liter. He has determined an athlete should drink 3/16 of a liter before each game. How many 3/16-liter servings can he pour from his test batch?

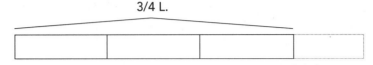

3/4 L.

Dr. Fumble can pour _____ 3/16-liter servings from his test batch.

Read aloud the problem. Have students circle the question and underline the relevant information on their papers. Ask: *What kind of bar model do we need to represent and solve this problem?* (Part-Whole, because we are given the whole and a part) Challenge students to work individually or in pairs to represent the problem with a bar model and solve it. Then call on volunteers to share their diagrams and strategies on the board.

Here is one way to approach this problem: Draw a bar representing a liter and divide it into four equal portions. The last 1/4 should be drawn with a dotted line to signify that our whole is equal to 3/4 of a liter. (See above.)

Using our knowledge of equivalent fractions, we can determine that 1/4 = 4/16. Draw lines to divide each of the fourths on the bar into 4 smaller sections, each representing 1/16 of a liter. Place the sixteenths into groups of 3, each representing one serving, as shown below.

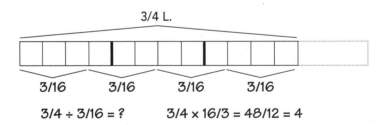

3/4 L.

3/16   3/16   3/16   3/16

$3/4 \div 3/16 = ?$ $3/4 \times 16/3 = 48/12 = 4$

By counting the number of 3/16-liter servings, we can see there are 4 servings in 3/4 liter. Show how this would also work with the standard "invert and multiply" approach: $3/4 \div 3/16 = 3/4 \times 16/3 = 4$.

Have students work on Problems #23 and 24 (page 67) independently or in pairs. Remind them to draw bars and write equations to represent each problem. Display each problem on the whiteboard, then call on volunteers to share their work and strategies on the board.

Decimal Problems

Materials: student pages 68–69, pencils, projector, interactive whiteboard, markers

Preparation: Distribute copies of pages 68–69 and pencils to students. Go to www.scholastic.com/problemsolvedgr6 and click on Lesson 7. Set up your computer and projector to display the problems on the interactive whiteboard.

Students who do not have a solid understanding of place value may struggle with decimals. It may be helpful to direct students to properly pronounce the word name of a decimal and not fall back on saying "point" when reading a decimal. For example, 0.41 should be pronounced as "forty-one hundredths" and not "point four-one." Understanding the value of the numbers with decimals in a word problem is essential to properly drawing bar models.

Display Problem #25 on the interactive whiteboard.

Max, Mabel, and Minky each have a collection of old Martian zloxny coins. Mabel has 2.5 times more coins than Max, and Minky has 1.5 times more coins than Mabel. If they have 638 zloxny coins in all, how many does Mabel have?

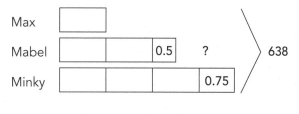

Mabel has _____ zloxny coins.

Read aloud the problem. Then have students circle the question and underline the relevant information on their papers. Ask the class: *What type of bar model does this problem call for?* (Comparison, because we are comparing three people's coin collections)

Explain that since Mabel's quantity is a multiple of Max's and Minky's quantity is a multiple of Mabel's, Max has the smallest quantity and will serve as the unit. If Max is 1 unit and Mabel is 2.5 times larger, then Mabel has $1 \times 2.5 = 2.5$ units. If Minky is 1.5 times larger than Mabel, then Minky has $1.5 \times 2.5 = 3.75$ units.

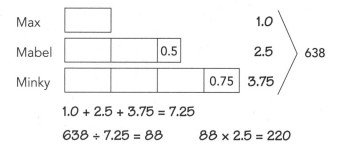

$$1.0 + 2.5 + 3.75 = 7.25$$
$$638 \div 7.25 = 88 \qquad 88 \times 2.5 = 220$$

Let's add up the number of units: 1 + 2.5 + 3.75 = 7.25 units. If we divide the total number of coins they have in all by the number of units, we get 638 ÷ 7.25 = 88 coins per unit. Since Mabel has 2.5 units, she has 2.5 × 88 = 220 coins.

Display Problem #26 on the interactive whiteboard.

Quikster and Speedy work at Harris's Hotcakes 'n' Haggis food truck. Quikster earns $10.50 more than Speedy each week. They each spend $50 per week on yummy school lunches and donate the rest to the school's triangle and kazoo marching band. When Quikster's donations totaled $300, Speedy's totaled $90. How much does Speedy earn each week?

Q.		$10.50

S.	

?

Speedy earns $____ each week.

Fractions and Decimals

As students work with fractions and decimals, consider creating with the class an equivalence chart for benchmark fractions and decimals, such as 1/2 = 0.50 and 1/4 = 0.25. Link these numerical expressions with their word forms as well: 3/4 = 0.75 = three fourths = seventy-five hundredths. Include diagrams, using graph paper (especially 10×10 grids) that can be colored in to show the amount being described.

Read aloud the problem. Have students circle the question and underline the relevant information on their papers. Explain to the class that this is another problem best represented by a Comparison bar model. Challenge students to work individually or in pairs to represent the problem with a bar model and solve it. Then call on volunteers to share their diagrams and strategies on the board.

Here is one way to approach this problem: First, set up a Comparison bar model depicting the difference between Quikster's and Speedy's weekly earnings. (See above.) Then, add information to show that they each spend $50 per week and donate the rest.

Quikster's donation is $10.50 more per week than Speedy's. After an unknown number of weeks, Quikster had donated $300 and Speedy had donated $90, a difference of $210. To find the number of weeks, divide the difference in total donations ($210) by the weekly difference in donations ($10.50): $210 ÷ $10.50 = 20 weeks. So if it took Speedy 20 weeks to donate $90, that means he donated $4.50 per week ($90 ÷ 20). And if he spends $50 per week and donates the rest, then Speedy makes $50 + $4.50 = $54.50 per week.

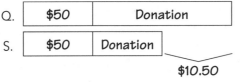

$300 – $90 = $210

$210 ÷ $10.50 = 20 weeks

$90 ÷ 20 = $4.50

$50 + $4.50 = $54.50

Have students work on Problems #27 and 28 (page 69) independently or in pairs. Remind them to draw bars and write equations to represent each problem. Display each problem on the whiteboard then call on volunteers to share their work and strategies on the board.

LESSON 8

Percentage Problems

Materials: student pages 70–71, pencils, projector, interactive whiteboard, markers

Preparation: Distribute copies of pages 70–71 and pencils to students. Go to www.scholastic.com/problemsolvedgr6 and click on Lesson 8. Set up your computer and projector to display the problems on the interactive whiteboard.

Although this chapter is divided into separate lessons for percentages, decimals, and fractions, emphasize to students that these three conventions are just different ways of showing the same thing, i.e., parts of a whole. Each form may be seen more frequently in different real-world contexts, and sometimes one form is easier to use than another in a particular calculation. Ask students, for example, where they have heard the term *percent* or *percentage* in real-world contexts. (Answers may include sales or reduction in list price, tips, sales tax rates, scores on exams, and so on.) Students should be comfortable using percentages, decimals, and fractions, and converting from one form to the others (e.g., 0.75 = 75% = 75/100 = 3/4).

Display Problem #29 on the interactive whiteboard.

Tina Trashtruckian, celebrity extraordinaire, ate lunch at her favorite Beverly Hills restaurant, Le Petite Snob. She wasn't going to leave a tip, but her agent persuaded her to, so she left $184.56. That included the cost of her meal plus a 20% tip. What was the cost of the meal before the tip?

?

Before Tip: | 20% | 20% | 20% | 20% | 20% |

The cost of the meal before the tip was $_____.

Read aloud the problem. Then have students circle the question and underline the relevant information on their papers.

Point out that if Tina gave a 20% tip that means she added an amount equal to 20% of the original restaurant bill. Since 100 can be split into five 20s, we can start there. (See above.)

When we draw the "after tip" bar, we should have an additional 20% section, giving us 6 sections. We can find the value of one segment by dividing $184.56 ÷ 6 = $30.76. If one segment is worth $30.76 and the "before tip" bar has 5 segments, then the value of the meal before tip was added is $184.56 – $30.76 = $153.80. We can check our work by taking the original amount, $153.80, and adding the tip of $30.76 to arrive at the total of $184.56.

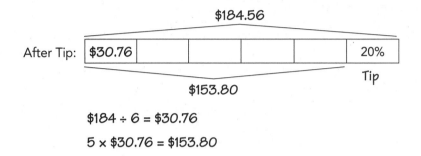

$$\$184 ÷ 6 = \$30.76$$
$$5 × \$30.76 = \$153.80$$

Display Problem #30 on the interactive whiteboard.

Alvin bought a bunch of fleas at the local flea market. He graciously gave 15% of them to his sister, Allie. If he gave 54 fleas to Allie, how many did he buy originally?

?

5%

54

Alvin originally bought _____ fleas.

Read aloud the problem. Have students circle the question and underline the relevant information on their papers. Then challenge students to work individually or in pairs to represent the problem with a bar model and solve it. Call on volunteers to share their diagrams and strategies for solving the problem on the board.

Here is one way to approach this problem: Start by drawing a bar to represent all the fleas Alvin bought. Since 15% is not a factor of 100%, we could divide the bar into 20 sections, each representing 5% of the number of fleas purchased. (See above.)

We know that 54 fleas represents 15% of the total, and 15% represents 3 out of 20 sections. To find 5% of the total, we can divide 54 ÷ 3 = 18. So if one section equals 18 fleas, then 20 sections equal 20 × 18 = 360 fleas.

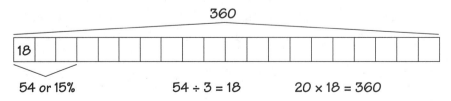

360

18

54 or 15%   54 ÷ 3 = 18   20 × 18 = 360

Have students work on Problems #31 and 32 (page 71) independently or in pairs. Remind them to draw bars and write equations to represent each problem. Display each problem on the whiteboard then call on volunteers to share their work and strategies on the board.

Fraction/Decimal/Percentage Challenge Problems

Materials: student pages 72–73, pencils, projector, interactive whiteboard, markers

Preparation: Distribute copies of pages 72–73 and pencils to students. Go to www.scholastic.com/problemsolvedgr6 and click on Lesson 9. Set up your computer and projector to display the problems on the interactive whiteboard.

Tell students that in this lesson they will encounter word problems with a mix of fractions, decimals, and percentages that will enable them to practice the skills and techniques they've learned thus far. Remind students to circle what the problem wants answered and underline relevant information to help them keep their facts straight.

Display Problem #33 on the interactive whiteboard.

Antiques aficionado Lulu recently appeared on *The Old Stuff Show*. The expert appraiser valued her vintage collection of disposable dishware, including 2 used Styrofoam cups, 2 mint-condition plastic forks, and an empty milk carton, at a total of $52,000. A fork's value is 50% as much as a cup, and the milk carton is valued at $3,000 more than a fork. What is the value of the milk carton?

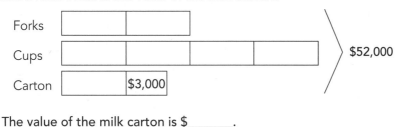

The value of the milk carton is $_____.

Read aloud the problem. Have students circle the question and underline the relevant information on their papers. Explain to the class that a Comparison model would be good here because the parts of the whole are given in amounts relative to one another.

Start with the fork as the unit because it is the smallest in value. Since we have 2 forks, the forks will be 2 units in length. Then *each* cup will be worth 2 units, because the problem states that a fork is worth 50% of a cup. Since there are 2 cups, the bar representing the cups will include 4 units. Finally, the bar for the milk carton will also include one unit but will also have another section worth $3,000, since the problem states that we have one milk carton and it is worth $3,000 more than a fork. (See the diagram on page 27.)

Looking at the bars in total, we have 7 units plus $3,000 equal to $52,000. That means 7 units equal $52,000 – $3,000 = $49,000. If 7 units have a value of $49,000, then each unit has a value of $49,000 ÷ 7 = $7,000. The value of the milk carton is $7,000 + $3,000 = $10,000.

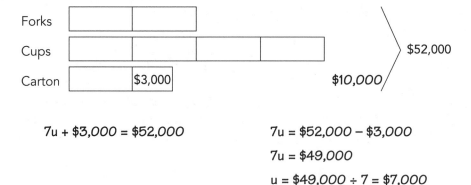

$$7u + \$3{,}000 = \$52{,}000$$

$$7u = \$52{,}000 - \$3{,}000$$
$$7u = \$49{,}000$$
$$u = \$49{,}000 \div 7 = \$7{,}000$$
$$\$7{,}000 + \$3{,}000 = \$10{,}000$$

((o)) **Display Problem #34 on the interactive whiteboard.**

Red bought some supplies for a trip to visit her grandmother. She spent 0.4 of her money on a picnic basket filled with snacks and 5/12 of the remainder on wolf repellent. The wolf repellent cost $30.00. How much money did Red start with?

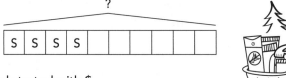

Red started with $_____.

Read aloud the problem. Have students circle the question and underline the relevant information on their papers. Explain that this problem calls for a Part-Whole bar model since we know the parts and are looking for the whole. Challenge students to work individually or in pairs to represent the problem with a bar model and solve it. Then call on volunteers to share their diagrams and solutions on the board.

Here is one way to approach the problem: The amount of money Red started with is the unknown, so draw a bar with an arrow and question mark to represent the total. We know that 0.4, or 4 out of 10 units, were spent on snacks. So let's split the bar into 10 sections and label 4 of them as snacks. (See above.)

Draw another bar to represent the 0.6 part of the bar not spent on snacks, and split it into 12 sections. If 5 of the sections are worth a total of $30, then each section is worth $6, and the entire second bar is worth $72 (12 × $6).

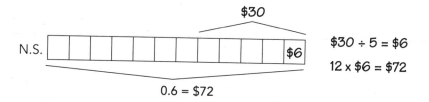

$30 ÷ 5 = $6

12 × $6 = $72

Going back to the original bar, we now know that 6 sections have a value of $72, so each section has a value of $12 ($72 ÷ 6). There are 10 sections in the original bar, so the amount of money Red started off with was $120 (10 × $12).

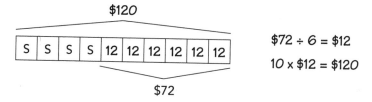

$72 ÷ 6 = $12

10 × $12 = $120

Have students work on Problems #35 and 36 (page 73) independently or in pairs. Remind them to draw bars and write equations to represent each problem. Display each problem on the whiteboard then call on volunteers to share their work and strategies on the board.

Using Calculators

Sometimes you may want to allow students to use calculators when solving problems. With some problems, you may want to emphasize the problem-solving process and figuring out what steps and strategies make sense in attacking the problem. The computations involved might be secondary on these occasions, and calculator use can help support that focus.

Bar Modeling With Ratios and Proportions

Ratios and Proportional Reasoning is a math domain that appears only in the 6th and 7th grade Common Core State Standards, highlighting the grades when students transition from arithmetic to algebra. In addition to this being a new concept, students have the added challenge of learning new notation. They know from their work in the lower grades that the fraction 3/4 means "three parts out of a whole consisting of four parts." A three-to-two ratio, on the other hand, can be written as 3:2, "three to two," or 3/2.

Before proceeding with the lessons in this chapter, make sure students are familiar with ratio notation so they can fully understand the directions and apply the Bar Modeling technique.

Ratio Basics

Materials: student pages 74–75, pencils, projector, interactive whiteboard, markers

Preparation: Distribute copies of pages 74–75 and pencils to students. Go to www.scholastic.com/problemsolvedgr6 and click on Lesson 10. Set up your computer and projector to display the problems on the interactive whiteboard.

In this lesson students will solve multistep problems involving various operations in ratio and proportion situations. Using more than one diagram for a problem will be a likely option. Many of the problems will require logical thinking, and the Bar Modeling process can help students determine viable strategies for these various problem types. When students are working on ratios there is usually little reason for equations. The bars and tables provide strategies and solutions.

Display Problem #37 on the interactive whiteboard.

Renata's recipe for her famous Kale 'n' Clam Quencher calls for kale juice and clam juice in a 3:1 ratio. If Renata pours 18 liters of fresh kale juice into a vat, how many liters of clam juice should she add?

	3	6	9	12	15	18	21	24	27
K.J.									

	1	2	3	4	5	6	7	8	9
C.J.									

Renata should add _____ liters of clam juice.

Read aloud the problem. Have students circle the question and underline the relevant information on their papers. Ask the class: *Does this problem call for a Part-Whole bar model or a Comparison bar model?* (Comparison, since we're comparing the quantities of kale juice and clam juice)

On the board, set up two bars to compare the quantity of kale juice to the quantity of clam juice. Make sure the segments in the two bars are lined up, and label the segments according to the ratio in the directions. In this case, the kale juice segment will increase by 3 while the clam juice segments increase by 1. (See the diagram on page 30.)

Find 18 on the kale juice bar and shade in the segments representing 18 liters. Then, shade in an equal number of segments on the clam juice bar.

	3	6	9	12	15	18	21	24	27
K.J.									

	1	2	3	4	5	6	7	8	9
C.J.									

By comparing the bars, we can see that Renata would need to add 6 liters of clam juice.

Display Problem #38 on the interactive whiteboard.

Little Clara has 20 pets in all. Some are warthogs, and some are cobras. If the ratio of cobras to warthogs is 7:3, how many cobras does Little Clara have?

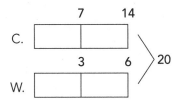

Little Clara has _____ cobras.

Read aloud the problem. Have students circle the question and underline the relevant information on their papers. Challenge students to work individually or in pairs to represent the problem with a bar model and solve it. Then call on volunteers to share their diagrams and strategies for solving the problem on the board.

Here is one way to approach the problem: Point out that in this problem, we know the ratio and the total, but not the parts. Set up two bars comparing the number of warthogs to the number of cobras. Make sure the segments are lined up, and label the quantity per segment for each critter, as shown above.

We can see that the label for 14 cobras lines up with 6 warthogs, and 14 + 6 = 20, so Clara has 14 cobras.

Labeling

Point out to students that they should label quantities in a diagram only when it is helpful. Don't make labeling a requirement for every problem. For example, if a problem compares quantities of carrots and peas, it makes sense to label each quantity to keep them straight. However, if a problem talks about a quantity of pencils, where some were added or taken away, there's no need to label that quantity since there is nothing to differentiate.

Have students work on Problems #39 and 40 (page 75) independently or in pairs. Remind them to draw bars to represent each problem. Give students a few minutes to work. Then display each problem on the whiteboard. Call on volunteers to draw and label the diagrams on the board. Then invite students to share their strategies for solving the problems, interacting on the whiteboard whenever possible.

LESSON 11

Ratios and Fractions

Materials: student pages 76–77, pencils, projector, interactive whiteboard, markers

Preparation: Distribute copies of pages 76–77 and pencils to students. Go to www.scholastic.com/problemsolvedgr6 and click on Lesson 11. Set up your computer and projector to display the problems on the interactive whiteboard.

Students should have a solid understanding of fractions and ratios separately before attempting word problems that include both fractions and ratios. It's important they understand that a fraction represents the relationship between a part and the whole and a ratio represents the relationship between a part and a part. But if your class is confusing the two, run through a couple of simple examples of converting fractions to ratios and ratios to fractions.

Pose the following introductory problem to your students: *A recipe for chocolate-covered blueberries calls for 2 ounces of blueberries and 1 ounce of chocolate. What is the ratio of blueberries to chocolate?* (2:1) *What is the ratio of chocolate to blueberries?* (1:2) *What fraction of the recipe consists of blueberries?* (2/3) *What fraction consists of chocolate?* (1/3)

Point out that there are 3 parts in the recipe—2 are blueberries and 1 is chocolate. Review how these parts are put together to make the ratios and fractions.

Display Problem #41 on the interactive whiteboard.

At the end of a hard day of interplanetary conquest, evil Galactic Emperor Zog relaxes with a tall glass of his special chocolate milk. He mixes 16 ounces of fine Belgian chocolate syrup with 4 ounces of whole milk. What is the ratio of chocolate syrup to milk? What fraction of the chocolate milk consists of milk?

C. [   ] 16

M. [   ] 4

The ratio of chocolate syrup to milk is _____.

The fraction of the milk in the chocolate milk is _____.

Read aloud the problem. Have students circle the question and underline the relevant information on their papers. Point out that the unit of measurement for both ingredients in the problem is in ounces.

On the board, draw two bars divided into the number of ounces for each ingredient, as shown on page 32.

To find the ratio of chocolate syrup to milk, count the number of segments for each ingredient and write them as a ratio (16:4). Explain that like fractions, it is preferable to reduce ratios to simplest terms: 16:4 reduces to 4:1.

To find the numerator of the fraction of the drink that consists of milk, count the number of milk segments (4). For the denominator, count the total number of segments (20). The fraction of the drink that consists of milk is 4/20, which equals 1/5 when reduced to simplest terms.

Display Problem #42 on the interactive whiteboard.

Milton and Sydney are the two leading scorers on Slobovia's Olympic Jacks Team. Milton has scored half the team's points, and Sydney has scored 1/6 of the team's points. What is the ratio of Milton's points to Sydney's points?

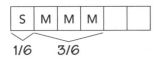

The ratio of Milton's points to Sydney's points is _____.

Read aloud the problem. Have students circle the question and underline the relevant information on their papers. Then challenge students to work individually or in pairs to represent the problem with a bar model and solve it. Call on volunteers to share their diagrams and strategies for solving the problem on the board.

Here is one way to approach this problem: We can draw a single bar to represent all the team's points. Sydney has scored 1/6 of the points, so we can divide the bar into 6 sections and label one section to represent his points. Milton has scored 1/2 of the team's points. That's equivalent to 3/6, so we can label 3 of the sections to represent his points. (See above.)

Since Sydney's points equal 1 segment and Milton's equal 3 segments, the ratio of Milton's points to Sydney's points is 3:1.

Have students work on Problems #43 and 44 (page 77) independently or in pairs. Remind them to draw bars to represent each problem. Give students a few minutes to work. Then display each problem on the whiteboard. Call on volunteers to draw and label the diagrams on the board. Then invite students to share their strategies for solving the problems, interacting on the whiteboard whenever possible.

Ratios With More Than Two Quantities

Materials: student pages 78–79, pencils, projector, interactive whiteboard, markers

Preparation: Distribute copies of pages 78–79 and pencils to students. Go to www.scholastic.com/problemsolvedgr6 and click on Lesson 12. Set up your computer and projector to display the problems on the interactive whiteboard.

Some students are concerned when they encounter ratios with more than two quantities. Going from two quantities to three in a ratio does add complexity to a problem, but the strategies they have learned so far are still applicable.

For example, if there is a 3:2 ratio of vanilla ice cream to hot fudge, we can also say that there is a 2:3 ratio of hot fudge to vanilla ice cream and that hot fudge is 2/5 of the sundae and vanilla ice cream is 3/5 of the sundae.

If we add, say, one unit of whipped cream so that the ratio of vanilla ice cream to hot fudge to whipped cream is 3:2:1, we can now determine six different ratios between any two individual ingredients, e.g., hot fudge to whipped cream is 2:1, whipped cream to vanilla ice cream is 1:3, and so on. We can also make three individual fractions: vanilla ice cream, hot fudge, and whipped cream are 1/2, 1/3, and 1/6 of the sundae, respectively.

Display Problem #45 on the interactive whiteboard.

Musicians Marvin, Millie, and Meriwether are raising money to fund the school banjo ensemble's overseas tour. Altogether they have received donations totaling $600. If the ratio of money raised by Marvin, Millie, and Meriwether is 7:5:3, how much did Meriwether raise?

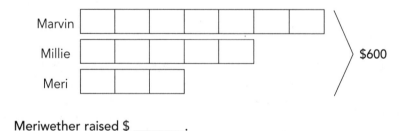

Meriwether raised $ _____.

Read aloud the problem. Have students circle the question and underline the relevant information on their papers.

Start by drawing three bars to compare the three musicians, with each bar having the number of units present in the ratio 7:5:3. The nice thing about ratios is that all the units will be the same size. (See above.)

We can see there are 15 units in all, and we know the total amount raised is $600. So the value of each unit is $600 ÷ 15 = $40. Since Meriwether's bar is 3 units long, he raised 3 × $40 = $120.

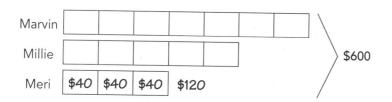

Marvin
Millie
Meri $40 | $40 | $40 | $120 $600

$600 ÷ 15 = $40   3 × $40 = $120

Display Problem #46 on the interactive whiteboard.

Selene's Gourmet Frozen Yogurt shop features "specialty" flavors. Yesterday, she sold servings of Mustard Ripple, Pepperoni, and Bacon in a ratio of 2:5:4. If she sold 36 servings of Mustard Ripple, how many servings did Selene sell in all?

M.R.   [|]   36

P.   [| | | |]

B.   [| | |]

Flavor of the Day Bacon

Selene sold _____ servings in all.

Read aloud the problem. Have students circle the question and underline the relevant information on their papers. Then challenge students to work individually or in pairs to represent the problem with a bar model and solve it. Call on volunteers to share their diagrams and strategies for solving the problem on the board.

Here is one way to approach this problem: Draw bars for each flavor. Each bar should have the number of units described in the ratio 2:5:4, as shown above.

If the bar for Mustard Ripple has 2 units and there were 36 servings sold, then each unit equals 18 servings (36 ÷ 2). To find how many servings were sold in all, count the total number of units (11) and multiply by the number of servings per unit (18): 11 × 18 = 198 servings.

M.R. | 18 | 18 | 36

P. | 18 | 18 | 18 | 18 | 18 | 36 ÷ 2 = 18

B. | 18 | 18 | 18 | 18 | 11 × 18 = 198

Have students work on Problems #47 and 48 (page 79) independently or in pairs. Remind them to draw bars to represent each problem. Give students a few minutes to work. Then display each problem on the whiteboard. Call on volunteers to draw and label the diagrams on the board. Then invite students to share their strategies for solving the problems, interacting on the whiteboard whenever possible.

Proportion Problems

Materials: student pages 80–81, pencils, projector, interactive whiteboard, markers

Preparation: Distribute copies of pages 80–81 and pencils to students. Go to www.scholastic.com/problemsolvedgr6 and click on Lesson 13. Set up your computer and projector to display the problems on the interactive whiteboard.

Proportion is another new topic for many 6th graders. While it may be new in terms of math instruction, kids have dealt with proportions their entire lives. Providing a few simple, real-world examples of how proportions are used (e.g., doubling a recipe) should help relieve anxiety.

Give the class a recipe for chocolate milk; for example, 1 ounce of chocolate and 7 ounces of milk. Ask how many ounces the recipe will yield. (*8 ounces*) Then ask what they would do if they were really thirsty, had a 16-ounce glass, and wanted to maintain the flavor from the original recipe. *(To maintain the same flavor, use 2 ounces of chocolate syrup and 14 ounces of milk to fill a 16-ounce glass.)* Explain that to keep the ingredients in proportion, they all have to be multiplied by the same factor. If necessary, show this in table form (see below).

Syrup (oz.)	1	2	3	4
Milk (oz.)	7	14	21	28
Size of Glass (oz.)	8	16	24	32

Display Problem #49 on the interactive whiteboard.

> Grandma Gertie's special birthday-cake icing recipe calls for 4 ounces of maple syrup and 5 ounces of blue cheese. If she decided to make a larger batch of icing and used 20 ounces of blue cheese, how many ounces of maple syrup would she need?
>
> M.S. [4]
>
> B.C. [5]
>
> Grandma Gertie would need ____ ounces of maple syrup.

Read aloud the problem. Have students circle the question and underline the relevant information on their papers.

Demonstrate on the board how to represent this problem with a bar model. Draw bars for each ingredient in the original recipe. Make the bar for blue cheese a bit longer than the bar for maple syrup. Label each bar with the number of ounces for that ingredient, as shown above.

If we increase the amount of blue cheese to 20 ounces, we would need to add 3 units of 5 ounces each to the bar.

B.C.	5	5	5	5

To keep the new batch in proportion, we would also have to add 3 units to the maple syrup bar.

B.C.	5	5	5	5

M.S.	4	4	4	4

Counting the number of units and ounces per unit, we determine that we would need 16 ounces (4 × 4) of maple syrup.

Display Problem #50 on the interactive whiteboard.

Lonesome Lenny Lewis, the town's most unpopular radio DJ, has decided to play only bagpipe tunes and accordion songs to appeal to the "young people." Each hour, he plays 6 bagpipe tunes and 4 accordion songs. Last night, he played 25 songs in all. How many featured the accordion?

B.

A.

___ songs featured the accordion.

Read aloud the problem. Have students circle the question and underline the relevant information on their papers. Then challenge students to work individually or in pairs to represent the problem with a bar model and solve it. Call on volunteers to share their diagrams and strategies for solving the problem on the board.

Here is one way to approach this problem: Start by representing the ratio of bagpipe tunes to accordion songs as 6:4. Note that this ratio can be reduced to simplest terms as 3:2. Represent this ratio with bars for each type of song, as shown below.

B.

A.

6:4 = 3:2

Now create a different bar to represent 25 songs (5 units of 5 songs each). If one block of 5 songs has 2 accordion songs, then 5 blocks would have 10 songs featuring the accordion (5 × 2).

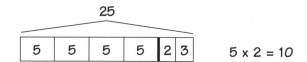

$5 \times 2 = 10$

Have students work on Problems #51 and 52 (page 81) independently or in pairs. Remind them to draw bars to represent each problem. Give students a few minutes to work. Then display each problem on the whiteboard. Call on volunteers to draw and label the diagrams on the board. Then invite students to share their strategies for solving the problems, interacting on the whiteboard whenever possible.

LESSON 14

Ratio and Proportion Challenge Problems

Materials: student pages 82–83, pencils, projector, interactive whiteboard, markers

Preparation: Distribute copies of pages 82–83 and pencils to students. Go to www.scholastic.com/problemsolvedgr6 and click on Lesson 14. Set up your computer and projector to display the problems on the interactive whiteboard.

This lesson features ratio and proportion concepts similar to those that students have already experienced, but in more complex situations.

Display Problem #53 on the interactive whiteboard.

Barry Sty, lead cowbell player in the band Pigpen, has a large collection of industrial vehicles that includes tanker trucks and cement mixers in a ratio of 4:5. He has 27 vehicles in all. If he decides to trade in 3 tanker trucks and buy 3 more cement mixers, what will the new ratio of tanker trucks to cement mixers be?

Before:

T.T. | 3 | 3 | 3 | 3

C.M. | 3 | 3 | 3 | 3 | 3

27

The new ratio of tanker trucks to cement mixers will be _____.

Read aloud the problem. Have students circle the question and underline the relevant information on their papers. Demonstrate how to represent this problem as a bar model.

We know that Barry has 27 vehicles in all. The ratio 4:5 means that we have 9 parts in all, so each part has a value of 3. This means Barry has 12 tanker trucks (4 × 3) and 15 cement mixers (5 × 3) for a total of 27 vehicles. Represent this with a "before" bar model, as shown on page 38.

If Barry trades in 3 tankers and buys 3 mixers, our representation would change to 3 units of 3 for tankers and 6 units of 3 for mixers. Share the diagram below, which represents the "after" portion of the problem.

After:

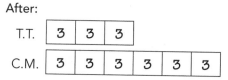

As we can see in the "after" diagram, the ratio of tankers to mixers is now 3:6, which reduces to 1:2 in simplest terms.

Display Problem #54 on the interactive whiteboard.

Rhonda and Miriam once had an equal number of pieces of pepper taffy. Rhonda ate 10 pieces, and Miriam ate 6 pieces. Now the ratio of Rhonda's taffy to Miriam's taffy is 2:3. How many pieces of pepper taffy did they each start with?

Before:

R.

M.

They each started with _____ pieces of pepper taffy.

Read aloud the problem. Have students circle the question and underline the relevant information on their papers. Then challenge students to work individually or in pairs to represent the problem with a bar model and solve it. Call on volunteers to share their diagrams and strategies for solving the problem on the board.

If needed, share this approach to solving the problem: Start by making a "before" representation of each girl's supply of taffy, as shown above.

As problems get more
complex, students may
need to be reminded
of the importance of
selecting a model type
that best fits the situation
depicted in the problem.
Let students know that
they may have to use
more than one model to
complete the problem.

If Rhonda ate 10 pieces and Miriam ate 6, then Miriam now has 4 more
pieces than Rhonda since they started with the same amount. Make an "after"
representation, with Rhonda's bar divided into 2 units and Miriam's bar divided
into 3 units, as dictated by the ratio information in the problem. Label the bars
to show that the difference between the two totals is 4 (see below).

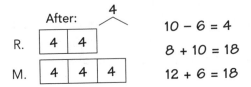

$$10 - 6 = 4$$
$$8 + 10 = 18$$
$$12 + 6 = 18$$

We can see that one unit has a value of 4, so Rhonda now has 8 pieces (2
units × 4) and Miriam has 12 (3 units × 4). If Rhonda ate 10 pieces and now
has 8, she started with 18 pieces. If Miriam ate 6 pieces and now has 12, she
also started with 18 pieces. So they each started with 18 pieces.

Have students work on Problems #55 and 56 (page 83) independently or in
pairs. Remind them to draw bars to represent each problem. Give students a
few minutes to work. Then display each problem on the whiteboard. Call on
volunteers to draw and label the diagrams on the board. Then invite students
to share their strategies for solving the problems, interacting on the whiteboard
whenever possible.

Bar Modeling With Algebra

Bar Modeling lends itself very well to helping students see how simple diagrams can be a means to setting up algebraic equations and developing strategies for solving them. In many ways, Bar Modeling assists students in representing their algebraic thinking as well as discovering it.

LESSON 15

Equation Basics

Materials: student pages 84–85, pencils, projector, interactive whiteboard, markers

Preparation: Distribute copies of pages 84–85 and pencils to students. Go to www.scholastic.com/problemsolvedgr6 and click on Lesson 15. Set up your computer and projector to display the problems on the interactive whiteboard.

Many students have a misconception about the meaning of the equal sign. To some, the sign means, "the answer to the problem is . . ." Teach students that the symbol means "the same as." In other words, what is on one side of the equation is the same as what is on the other side, in terms of value. If students master this conceptual understanding at the beginning of their study of algebra, they should have greater success when they move on to more advanced ideas, such as manipulating equations.

Write a simple equation on the board; for example, $5 + 3 = 8$. Ask students what the equal sign means. Some students may reply that it means, "the answer is" or "the sum is." Make the point that it means "the same as." It may be helpful to draw a balance scale on the board with $5 + 3$ on one arm and 8 on the other. (See below.) Demonstrate how adding 2 to both sides doesn't change the state of balance.

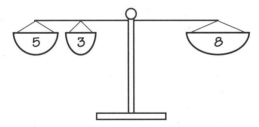

Write the algebraic equation $x + 3 = 8$ on the board. Point out that while $5 + 3 = 8$ is an equation, $x + 3 = 8$ is an *algebraic* equation because it includes the variable x, while $5 + 3 = 8$ does not. If necessary, provide a simple definition of *variable* as "a symbol for a number whose value we don't know yet."

Lester the forest gnome has a total of 20 pet toadstools,
which is 7 more than what his sister Lexie has.
How many pet toadstools does Lexie have?

| Lester | | 7 | 20 |

| Lexie | N |

Lexie has ____ pet toadstools.

Read aloud the problem. Have students circle the question (how many pet toadstools Lexie has) and underline the relevant information (Lester has 20 toadstools, 7 more than Lexie) on their papers.

Ask: *Would it be better to approach the problem with a Part-Whole bar model or a Comparison bar model?* (Comparison, since we are given one quantity in terms of the other)

To demonstrate how to represent and solve this problem, start with a bar to represent Lester's total, which is 20. Then draw a smaller bar underneath Lester's bar to represent Lexie's total. Label Lexie's bar with an *N* to indicate that the amount is unknown. Draw an arrow showing the difference between Lester's total and Lexie's total. Label the line with a 7.

| Lester | | 7 | 20 |

| Lexie | N |

$$N = 20 - 7 \quad N = 13$$

The equation is: $N = 20 - 7$, so $N = 13$. Lexie has 13 pet toadstools.

Maynard Mountain, quarterback for the New Jersey Noodles, owns 17 antique umbrellas. Nick Nicknack, the wide receiver, has 10 more than twice the number of umbrellas Maynard owns. How many umbrellas does Nick own?

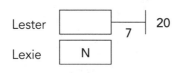

| Ni. | | | 10 | N |

| M. | 17 |

Nick owns _____ umbrellas.

Read aloud the problem. Have students circle the question and underline the relevant information on their papers. Then challenge students to work individually or in pairs to represent the problem with a bar model and solve it. Call on volunteers to share their diagrams and strategies for solving the problem on the board.

Here is one way to approach this problem: Draw a bar representing Maynard's 17 umbrellas. Above it, draw a bar with two sections, each section equal in length to Maynard's bar. To the right of the upper bar, draw a line representing the 10 additional umbrellas owned by Nick. (See the diagram on page 42.)

The equation is: $N = 2 \times 17 + 10$. $N = 44$. Nick owns 44 umbrellas.

Have students work on Problems #59 and 60 (page 85) independently or in pairs. Remind them to draw bars and write equations to represent each problem. Give students a few minutes to work. Then display each problem on the whiteboard. Call on volunteers to draw the diagrams and write equations on the board. Then invite students to share their strategies for solving the problems, interacting on the whiteboard whenever possible.

LESSON 16

Solving Equations

Materials: student pages 86–87, pencils, projector, interactive whiteboard, markers

Preparation: Distribute copies of pages 86–87 and pencils to students. Go to www.scholastic.com/problemsolvedgr6 and click on Lesson 16. Set up your computer and projector to display the problems on the interactive whiteboard.

If all equations given to students conveniently had the variable isolated on one side of the equation and everything else on the other, algebra would be a whole lot easier. The reality is that students must learn to manipulate equations, and we as teachers should remind them of the "11th Mathematical Commandment," which states: *What thou doest to one side of the equation, thou shalt do to the other.*

Display Problem #61 on the interactive whiteboard.

Arsene was transporting 5 containers of yummy turnip pie to be sold at his pie shop. Each container held 12 pies. While watching his favorite cartoon, *Super Gnat*, Arsene accidentally dropped a container, ruining 2/3 of the pies in the container. How many saleable pies were left?

12	12	12	12	12

There were _____ saleable pies left.

Read aloud the problem. Have students circle the question and underline the relevant information on their papers.

Set up a Part-Whole model since we know the number of parts and the value of each part. Start by drawing a bar with 4 sections to represent the 4 undamaged containers and another bar with 12 sections to represent the damaged container. (See below.) Divide the bar with 12 units into thirds, shading in the last 2/3 to represent the damaged pies. By adding the quantities in the two bars, we see that there are 52 undamaged pies.

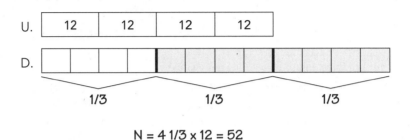

$$N = 4 \; 1/3 \times 12 = 52$$

The equation we need to solve is: $N = 4 \times 12 + 1/3 \times 12$. We can simplify the equation this way: $N = 4 \; 1/3 \times 12$. $N = 52$

Display Problem #62 on the interactive whiteboard.

Percy Pig was getting ready for a "visit" from B.B. Wolf. He ordered 4 truckloads of sticks to build his house out of flimsy, but easy-to-assemble sticks. His more practical brother, who had advised him to use brick, took pity on him and gave him 72 additional sticks. Percy now has 552 sticks. How many sticks were in each truckload?

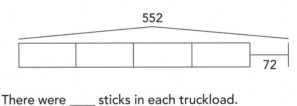

There were _____ sticks in each truckload.

Read aloud the problem. Have students circle the question and underline the relevant information on their papers. Then challenge students to work individually or in pairs to represent the problem with a bar model and solve it. Call on volunteers to share their diagrams and strategies for solving the problem on the board.

If needed, share this strategy for solving the problem: A Part-Whole model is appropriate since we know the whole, the number of parts, and the value of one of the parts. Draw a bar model with four equal parts and a fifth part representing the number of sticks donated by the practical brother, as shown above.

Put in equation form, the problem looks like this: 4T + 72 = 552.

Point out that when we work with equations, we want to have the variable—in this case, *T*—isolated on one side of the equal sign and everything else on the other side, because our objective is to find the value of the variable. In this problem, we can see that 4T + 72 = 552. To isolate T on one side, we first subtract 72 from each side of the equation. (Have students recall the example of the balance scales. If we have equal weights on each arm of the balance and remove an equal amount from each side, the scale stays in balance. The same is true with equations. That's why subtracting 72 from each side keeps the equation in balance.) When we subtract 72 from each side, we are left with 4T = 480. The next step is to divide both sides by 4 (again, keeping both sides of the equation balanced): T = 480 ÷ 4. T = 120. Each truckload carried 120 sticks.

Have students work on Problems #63 and 64 (page 87) independently or in pairs. Remind them to draw bars and write equations to represent each problem. Give students a few minutes to work. Then display each problem on the whiteboard. Call on volunteers to draw the diagrams and write equations on the board. Then invite students to share their strategies for solving the problems, interacting on the whiteboard whenever possible.

LESSON 17

Multistep Problems

Materials: student pages 88–89, pencils, projector, interactive whiteboard, markers

Preparation: Distribute copies of pages 88–89 and pencils to students. Go to www.scholastic.com/problemsolvedgr6 and click on Lesson 17. Set up your computer and projector to display the problems on the interactive whiteboard.

When working on multistep problems, students sometimes stop working after they solve one of the intermediate questions, believing they have finished the problem. Remind them to keep their focus on what the problem is asking to be solved (what they circle when they read the directions). Tell students that after they have completed work on a problem, it's a good practice to go back and ask, "Did I really answer the question the problem was asking?"

Display Problem #65 on the interactive whiteboard.

Lucretia went to a local craft fair to buy her mother birthday presents. She looked at a macaroni plaque, a garlic-scented bag of sachet, and a statue of a polar bear carved from a giant marshmallow. The 3 items cost $104 in all. The plaque cost $14 more than the sachet, and the sachet cost twice as much as the statue. In the end, she decided to buy 2 plaques and 2 statues. How much did she spend?

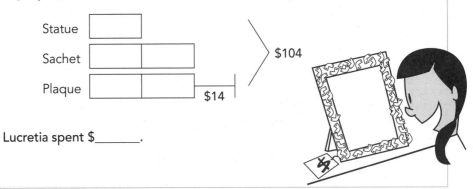

Lucretia spent $_____.

Read aloud the problem. Have students circle the question and underline the relevant information on their papers.

Demonstrate how to tackle this problem by setting up the following bar model: Draw bars to compare the cost of the three items, as shown above. Explain that since the statue has the lowest cost of the three items, it will serve as the unit. The sachet costs twice as much as the statue, so we draw two units to represent it. And the plaque costs $14 more than the sachet, so that's two units plus $14. Altogether, the three items cost $104.

So in all, we see that 5 units + $14 = $104. Subtracting $14 from both sides of the equation leaves 5 units = $90, so each unit equals $18. Using the bar model, the statue is 1 unit or $18, the sachet is 2 units or $36, and the plaque is 2 units plus $14 or $50. So if Lucretia buys 2 plaques and 2 statues, she would spend 2 × $50 + 2 × $18 = $136. Another way to solve the problem is to see that 2 statues would be 2 units and 2 plaques would be 4 units plus $28. 6 units plus $28 also equals $136.

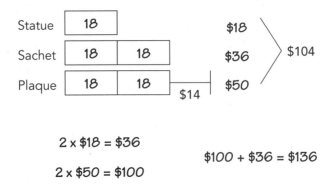

Display Problem #66 on the interactive whiteboard.

There are 360 students in the Pretend Creatures Club. For every 5 members who love unicorns, there are 7 members who fancy trolls. How many more troll lovers are there than unicorn lovers?

U. ☐☐☐☐☐
T. ☐☐☐☐☐☐☐ ⟩ 360

There are _____ more troll lovers than unicorn lovers.

Read aloud the problem. Have students circle the question and underline the relevant information on their papers. Then challenge students to work individually or in pairs to represent the problem with a bar model and solve it. Call on volunteers to share their diagrams and strategies for solving the problem on the board.

Here is one way to approach this problem: Compare the quantities using two bars, as shown in the diagram above. Set up this way, we have 12 units and 360 club members in all. That means each unit represents $360 \div 12 = 30$ people. Since there are 2 more units representing trolls than unicorns, there are 60 more troll fans (2×30). Another way to solve the problem is to find the number of unicorn lovers ($5 \times 30 = 150$) and subtract it from the number of troll fans ($7 \times 30 = 210$): $210 - 150 = 60$.

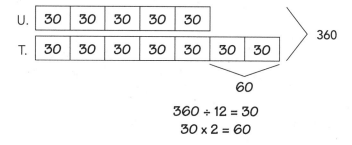

$$360 \div 12 = 30$$
$$30 \times 2 = 60$$

Have students work on Problems #67 and 68 (page 89) independently or in pairs. Remind them to draw bars and write equations to represent each problem. Give students a few minutes to work. Then display each problem on the whiteboard. Call on volunteers to draw the diagrams and write equations on the board. Then invite students to share their strategies for solving the problems, interacting on the whiteboard whenever possible.

Algebra Challenge Problems (Part I)

Materials: student pages 90–91, pencils, projector, interactive whiteboard, markers

Preparation: Distribute copies of pages 90–91 and pencils to students. Go to www.scholastic.com/problemsolvedgr6 and click on Lesson 18. Set up your computer and projector to display the problems on the interactive whiteboard.

As problems become more complex, there is often a need for more than one diagram to keep track of the various operations and strategies that are used to solve. These diagrams serve not only as a means of solving, but also as a record and proof of students' thinking about the problem.

Display Problem #69 on the interactive whiteboard.

Bumbling magician Marvin the Marvelous was having trouble with his pull-the-badger-out-of-a-hat trick, so he decided to practice it 100 times per day. On Monday, he was successful 4 times; on Tuesday, 12 times; on Wednesday, 19 times; and on Thursday, 6 times. If his overall success rate was 10% for a 5-day period, how many times was Marvin successful on Friday?

M	T	W	TH	F
100	100	100	100	100

Marvin was successful _____ times on Friday.

Read aloud the problem. Have students circle the question and underline the relevant information on their papers.

Guide students through the steps of drawing a bar model for this problem and solving it: First, draw a bar to represent the attempts for the week. Marvin practiced 100 times a day over a 5-day period. (See above.)

Then show how many successful attempts it would take on one of the days to achieve a 10% success rate. (See below.)

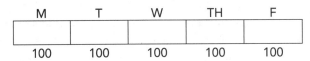

If 10 successful attempts out of 100 is necessary for one day, then 50 successful attempts would be needed for the 5-day period. Draw another bar to show the number of successful attempts each day. (See below.)

	50			
M	T	W	TH	F
4	12	19	6	N

$41 + N = 50$

$N = 9$

When we add up the sections that show the number of successful attempts each day, we get $41 + N = 50$. Solving for N, we determine that Marvin made 9 successful attempts on Friday.

Display Problem #70 on the interactive whiteboard.

The athletic director at Count Dracula Middle School ordered an equal number of uniforms for each of the school's three sports: swimming, gymnastics, and country-line dancing. There are 147 athletes in all. To get an equal number of athletes on each team, she moved 14 swimmers to gymnastics, 18 gymnasts to country-line dancing, and 12 dancers to swimming. How many athletes were on each team before the big switch?

	147	
Swimming	Gymnastics	Line Dancing

The starting number of swimmers was _____.

The starting number of gymnasts was _____.

The starting number of dancers was _____.

Read aloud the problem. Have students circle the question and underline the relevant information on their papers. Then challenge students to work individually or in pairs to represent the problem with bar models and solve it. Call on volunteers to share their diagrams and strategies for solving the problem on the board.

Here is one way to approach this problem: Since there are 147 athletes assigned to 3 teams with an equal number of players at the end of the year, we can determine the ending size of each team. (See above.) Each team ends with $147 \div 3 = 49$ athletes.

Let's draw a bar to represent each movement of students from beginning to end, starting with the swimmers:

Swimming

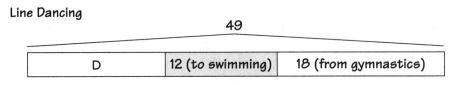

$$S - 14 + 12 = 49.$$ The starting number of swimmers was 51. (See above.)
Next, let's tackle the gymnasts:

Gymnastics

49

| G | 18 (to dancing) | 14 (from swimming) |

$$G - 18 + 14 = 49.$$ The starting number of gymnasts was 53. (See above.)
And, finally, the country-line dancers:

Line Dancing

49

| D | 12 (to swimming) | 18 (from gymnastics) |

$$D - 12 + 18 = 49.$$ The starting number of dancers was 43. (See above.)

Have students work on Problems #71 and 72 (page 91) independently or in pairs. Remind them to draw bars and write equations to represent each problem. Give students a few minutes to work. Then display each problem on the whiteboard. Call on volunteers to draw the diagrams and write equations on the board. Then invite students to share their strategies for solving the problems, interacting on the whiteboard whenever possible.

LESSON 19

Algebra Challenge Problems (Part II)

Materials: student pages 92–93, pencils, projector, interactive whiteboard, markers

Preparation: Distribute copies of pages 92–93 and pencils to students. Go to www.scholastic.com/problemsolvedgr6 and click on Lesson 19. Set up your computer and projector to display the problems on the interactive whiteboard.

Bar Modeling can help students keep track, not only of the various quantities in a problem, but also of how they are related to one another.

Display Problem #73 on the interactive whiteboard.

Tina Trashtruckian started a new fashion trend—toting around miniature squirrels in a bag. For a total of $1,310, she bought 3 Nantucket red squirrels, 2 Princetonian black squirrels, and 1 regular old gray squirrel. A red squirrel cost 3 times as much as a black squirrel, and the gray squirrel cost $50 more than a black squirrel. How much did the gray squirrel cost?

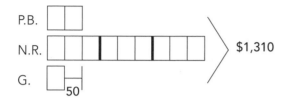

The gray squirrel cost $_____.

Read aloud the problem. Have students circle the question and underline the relevant information on their papers. Set up a Comparison bar model to demonstrate how this problem can be solved.

Draw three bars for the three types of squirrels, as shown above. Since the Princetonian black squirrel cost the least, it will serve as the unit. To show the number of squirrels Tina bought, draw 2 units for the 2 black squirrels, 9 units for the 3 red squirrels (since each red squirrel cost 3 times as much as a black squirrel), and 1 unit plus $50 for the gray squirrel (because the gray squirrel costs $50 more than a black squirrel).

In equation form, this looks like: 12 units + $50 = $1,310. Thus 12 units = 1,260. 1,260 ÷ 12 = 105, so a gray squirrel costs $105 + $50 = $155.

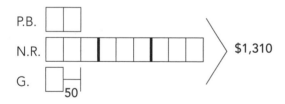

12 units + 50 = $1,310

12 units = $1,260

1 unit = $105

G = $105 + $50 = $155

Rufus can't wait for lunchtime. Today is quiche day! The cafeteria made 3 times as many beet quiches as squid-ink quiches and 50 more kelp quiches as squid-ink quiches. They made 945 quiches in all. How many more beet quiches are there than kelp quiches?

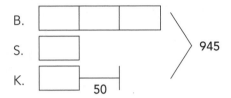

There are _____ more beet quiches than kelp quiches.

Silent Modeling

One fun and effective method of sharing a strategy is to tell students: *I am going to model my thinking, but I am not going to say anything. See if you can figure out what I am doing.* Do each step slowly and very deliberately on the board. Pause after each move and stand aside for a moment. At the end of your demonstration, ask students to share what they think you were doing.

Read aloud the problem. Have students circle the question and underline the relevant information on their papers. Then challenge students to work individually or with a partner to represent the problem with a bar model and solve it. Call on volunteers to share their diagrams and strategies for solving the problem on the board.

Here is one way to approach this problem: Make a Comparison model for the 3 flavors, as shown above. Note that we're using the squid-ink quiche as the primary unit, since that has the lowest quantity. So we draw one unit of squid-ink quiche, 3 units of beet quiche, and one unit of kelp quiche plus 50. Altogether, they add up to 945 quiches.

In equation form, we have: 5 units + 50 = 945. Subtracting 50 from both sides, we see that 5 units = 895. 895 ÷ 5 = 179 quiches per unit. Beets represents 3 units, so there are 3 × 179 = 537 beet quiches. Kelp represents 1 unit + 50, so there are 229 kelp quiches. The difference between beet and kelp quiches: 537 – 229 = 308. There are 308 more beet quiches than kelp quiches.

B.	179	179	179
S.	179
K.	179	50

945

5 units + 50 = 945

5 units = 895

1 unit = 179

B: 3 x 179 = 537   K: 179 + 50 = 229   537 – 229 = 308

Have students work on Problems #75 and 76 (page 93) independently or in pairs. Remind them to draw bars and write equations to represent each problem. Give students a few minutes to work. Then display each problem on the whiteboard. Call on volunteers to draw the diagrams and write equations on the board. Then invite students to share their strategies for solving the problems, interacting on the whiteboard whenever possible.

Bar Modeling With Distance, Rate, and Time

In 6th grade, students transition from arithmetic to algebra and begin to use formulas in their math and science classes. As they are introduced to ratio and proportion, they will become acquainted with the concept of unit rate. Bar Modeling can be helpful to students as they master these topics.

LESSON 20

Distance, Rate, and Time Problems

Materials: student pages 94–95, pencils, projector, interactive whiteboard, markers

Preparation: Distribute copies of pages 94–95 and pencils to students. Go to www.scholastic.com/problemsolvedgr6 and click on Lesson 20. Set up your computer and projector to display the problems on the interactive whiteboard.

Ask students where they might have heard the word *per* in their everyday life. Examples might include speed limits (miles per hour), supermarket prices (dollars per pound), and automobile mileage (miles per gallon).

Use the example of a car traveling at 40 miles per hour. Ask: *How many miles will the car travel in 1/2 hour?* (20 miles) *In 2 hours?* (80 miles) Then ask about a situation where the rate is unknown. For example: *If a car travels 90 miles in 2 hours, what is its average rate?* (45 miles per hour) Finally, ask a question where time is the unknown: *How long would it take a car traveling 30 mph to travel 90 miles?* (3 hours)

If appropriate for your class, introduce the formula $d = r \times t$ (distance = rate × time) and show how it works using the above examples.

Display Problem #77 on the interactive whiteboard.

Kyra was excited to win a place in this year's *R Games*. (*R* stands for Ridiculous!) In her medley event, she hopped on a pogo stick for 15 minutes at a rate of 8 km per hour and then rode her big-wheel tricycle for 1/2 hour at a rate of 6 km per hour. How far did she travel?

Kyra traveled _____ km.

By now, students have been exposed to lots of different problem types and operations. A good assignment is to have students create word problems for the class following certain criteria that you assign. For example: *Prepare a word problem that uses three 3-digit addends that will make us add and regroup.* Share each student's problem on the board, giving the author credit, and invite that student to share his or her strategy and solution. This is a great way to provide interesting practice problems while encouraging students to get creative with math.

Read aloud the problem. Ask students to circle the question (how far did she travel) on their papers. Then have them underline the relevant information (15 minutes at 8 km per hour and 1/2 hour at 6 km per hour).

One option to approach this problem is to draw a bar divided into 8 sections to depict the unit rate for the pogo stick, which is 8 km per hour. Then we divide the bar into 4 sections, each representing 1/4 of an hour, or 15 minutes, as shown below.

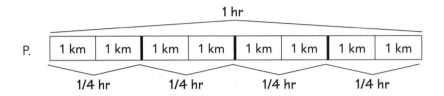

So we can see that Kyra traveled 2 km on the pogo stick portion of the event. Repeat the same process for the tricycle event, this time dividing the bar into 6 sections, depicting a rate of 6 km per hour, as shown below.

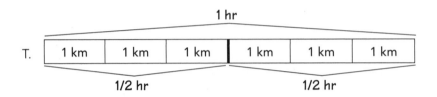

After dividing the bar into 2 sections, each representing 1/2 an hour, we can see that Kyra traveled 3 km on the tricycle. So 2 km + 3 km = 5 km in all.

Display Problem #78 on the interactive whiteboard.

Cassandra jet skied from Rotting Fish Beach to Polluted Point. She went 2/5 of the way in 2 hours. Then she shifted into high gear and went the rest of the way in 2 hours at 60 miles per hour. What was the average rate of speed for the whole trip?

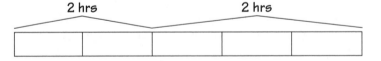

The average speed for the whole trip was _____ miles per hour.

Read aloud the problem. Have students circle the question (Cassandra's average rate of speed) and underline the relevant information (2/5 of the trip in 2 hours, the rest of the trip in 2 hours at 60 mph) on their papers. Then challenge students to work individually or in pairs to represent the problem with a bar model and solve it. Call on volunteers to share their diagrams and strategies for solving the problem on the board.

If needed, share this strategy with students: One approach is to set up a bar to show the distance for the trip. Divide the bar into 5 sections because the problem provides information in fifths. Also, show that 2/5 of the trip took 2 hours and 3/5 of the trip—the difference between 1 and 2/5—also took 2 hours. (See the diagram on page 54.)

Then set up a separate bar to show the 3/5 section of the trip, since we have information about rate and time, as shown below. Dividing the 120 miles for this part of the trip by 3 segments tells us that each segment covered 40 miles.

120 miles (2 hrs @ 60 mph)

40 miles	40 miles	40 miles
1/5	1/5	1/5

Then, go back to the first bar and write 40 in each section. In all, it took 4 hours to jet ski 200 miles: 200 ÷ 4 = 50 mph.

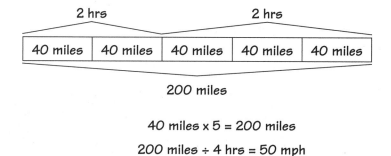

2 hrs   2 hrs

40 miles	40 miles	40 miles	40 miles	40 miles

200 miles

40 miles x 5 = 200 miles

200 miles ÷ 4 hrs = 50 mph

Have students work on Problems #79 and 80 (page 95) independently or in pairs. Remind them to draw bars and write equations to represent each problem. Give students a few minutes to work. Then display each problem on the whiteboard. Call on volunteers to draw the diagrams and write equations on the board. Then invite students to share their strategies for solving the problems, interacting on the whiteboard whenever possible.

Name _Gracie_ ♡

1. Prunella was admiring her dental floss collection. She has 207 pieces of mint-flavored floss, 1,194 pieces of cinnamon-flavored floss, and 43 pieces of extremely rare kumquat-flavored floss. How many pieces of dental floss does Prunella have in all?

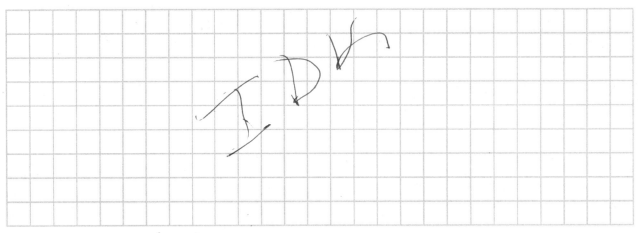

Prunella has _1444_ pieces of dental floss in all.

2. It's the first day of business for Francine's Fez shop. Prior to the big day, Francine worked for 5 days in a row, 18 hours each day, making high-quality fezzes. By the time the doors opened, she had made 947 fezzes. After serving a happy throng of customers, she had only 159 fezzes left. How many fezzes did she sell on her first day of business?

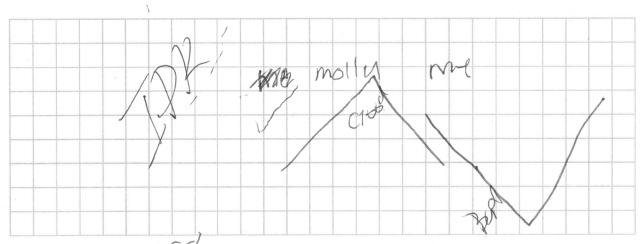

Francine sold _788_ fezzes on her first day of business.

Name _Grant_

3. Maurice Morris and his Morris Dancers are the hottest new club act in Funkyville. People just can't get enough of the 16th-century English folk sound! At a recent two-night engagement at Club Poseur, 2,417 customers came to see the act. If 1,283 customers caught his show the first night, how many saw the act the second night?

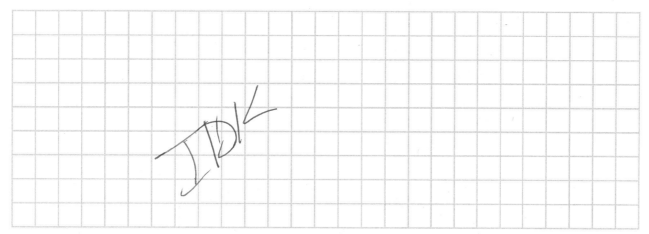

_1,134_ customers saw the show on the second night.

4. Harris Hambone has finally saved up enough money to realize his dream of opening his own food truck, which he's decided to call Harris's Hotcakes 'n' Haggis. On Monday, Harris sold 458 orders. On Tuesday, he sold 166 more orders than he sold on Monday. How many orders did he sell on Tuesday?

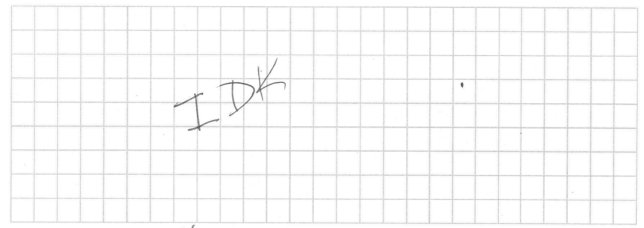

Harris sold _624_ orders on Tuesday.

Name _Gracie ♡_

5. Celebrity Tina Trashtruckian's living room has 27 mirrors so she can always see how she looks from any angle. The room also has 3 times as many pictures of the person Tina loves most in this world—herself—as it does mirrors. How many pictures of Tina are in the living room?

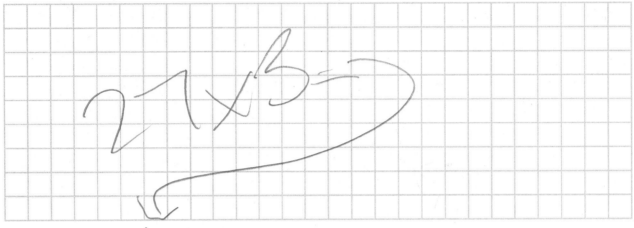

There are __81__ pictures of Tina in the living room.

6. Estelle's secret recipe for caramel beet dainties calls for 39 beets. For the coming holidays, she's going to make a batch and split it among 3 friends. How many beets will be in the gift each friend receives?

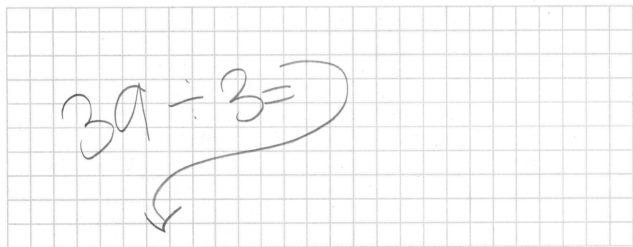

There will be __13__ beets in each gift.

Name _____

7. Daphne landed the job of her dreams, grooming Mandrake, the pet warthog of celebrity Tina Trashtruckian. In return for painting Mandrake's nails, polishing his chrome collar, and brushing his tusks, Daphne gets paid a sweet $27 per week. After 7 weeks, how much did Daphne make in all?

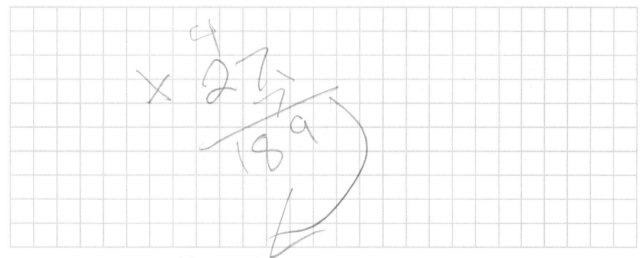

Daphne made $ ___180___ in all.

- -

8. Rocco Rat has his heart set on buying the deluxe edition of *The Diary of a Suburban Chipmunk*, which costs $256. Rocco is able to save $32 per month for the book. How many months will it take him to save enough to have this treasure for his very own?

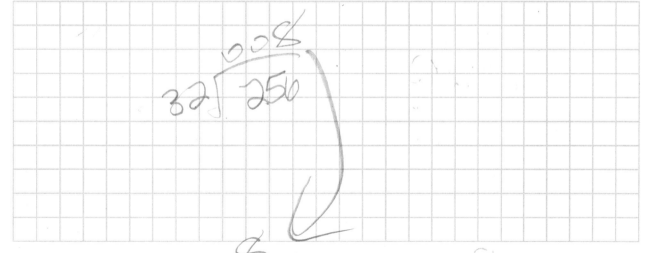

Rocco needs to save for ___8___ months.

Name _____

9. Clementine had 18 pet wolverines. Her mother insisted that she give them away. Clementine had no choice but to obey and donate them to nature centers. If each nature center will receive 1/6 of Clementine's pack, how many wolverines will go to each center?

Each center will receive ___3___ wolverines.

10. Petunia had $280. To make a powerful fashion statement, she spent 5/7 of her money on a pioneer-style sunbonnet, created by celebrity designer Dieter. After paying for the sunbonnet, how much money did Petunia have left?

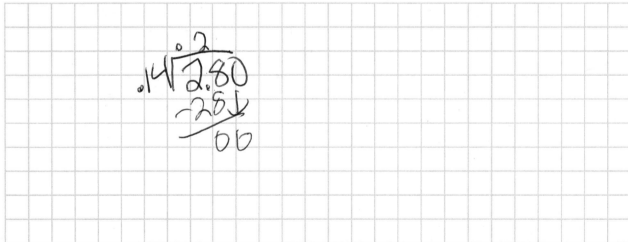

Petunia had $ ___20___ left.

Name _____

11. Chauncey bought a bag of 366 gumballs. 2/3 of them were habanero-pepper flavored, and the rest were cheesesteak flavored. How many cheesesteak-flavored gumballs did Chauncey buy?

Chauncey bought __5 6__ cheesesteak-flavored gumballs.

12. Teenage celebrity Dustin Dweeber views himself as a legend. To celebrate his greatness, he bought 35 crowns to wear when he meets his fans. 3/7 of the crowns are fool's gold, and the rest are aluminum. How many are aluminum?

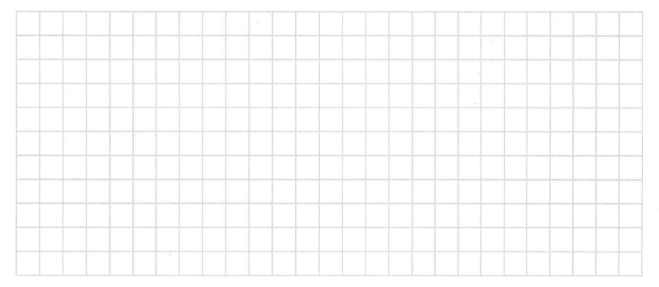

Dustin has _____ aluminum crowns.

13. Manfred had his heart set on buying the bestseller, *Everything I Needed to Know, I Learned From My Gerbil.* He spent 3/10 of his savings on the book, which cost $21.21. How much did he have in savings before he made the purchase?

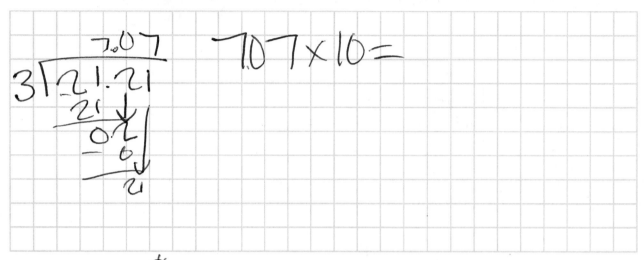

$$7.07 \times 10 =$$

Manfred had $ 70.7 in savings before he made the purchase.

14. 4,200 people auditioned for the new hit TV show, *I'm Dying to Be on Television!* 2/7 → 1200 of the people in line were clumsy jugglers, 8/14 of the contestants were off-key singers, and the rest were humorless comedians. How many were humorless comedians?

2400 ↑

There were _600_ humorless comedians.

15. Chef Marco spent 1/7 of his money on an atomic prune pitter. He used 2/3 of what was left to set up a bank account for his pet cockroach, Ibiza. After that, he had $5.48 left. How much money did Chef Marco have before he bought the prune pitter?

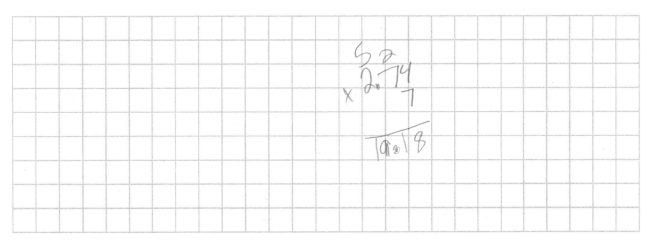

Chef Marco had $ __19.18__ before he bought the prune pitter.

16. Professor Lincoln Talbot IV has a little-known fascination with the My Tiny Horsey series of children's books. 1/3 of his collection features the beloved character Anchovy, 5/6 of the remainder of the collection are from the Comb My Mane collection, and 3 books are Oats 'n' Apples special editions. How many books featuring Anchovy does Professor Talbot own?

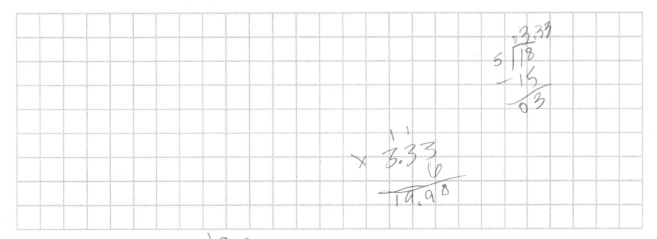

Professor Talbot owns __9.98__ books featuring Anchovy.

17. Gilbert's father ordered him to dispose of his overly ripe, smelly cheese collection. He gave 2/5 of his collection to the Hold-Your-Nose Cheese Emporium and 1/3 of the collection to the school cafeteria. If the Cheese Emporium received 30 more pieces of cheese than the cafeteria, how many pieces of cheese did the cafeteria get?

$$\begin{array}{r} 90 \\ 1\overline{\smash{\big)}90} \end{array}$$

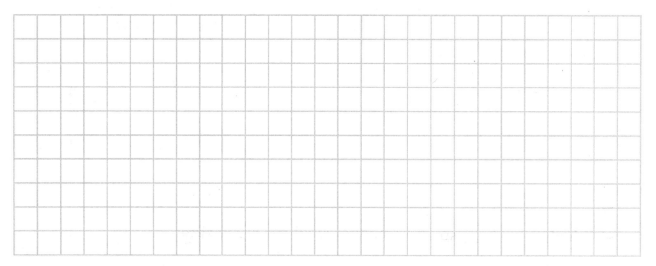

The cafeteria got __120__ pieces of cheese.

18. Wilhelm the worm was comparing his length to that of his siblings. Wolfric is 1/3 of Wilhelm's length, and Waldo is twice as long as Wilhelm. If Wolfric is 12 centimeters long, how long is Waldo?

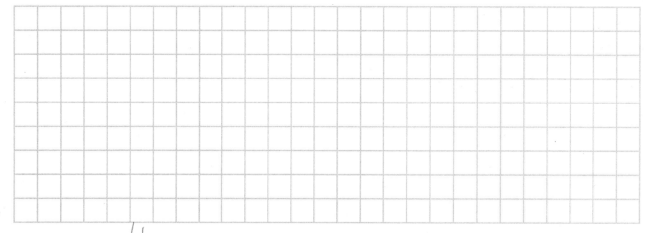

Waldo is __4__ centimeters long.

Name _____

19. Boy-Genius Leonard blew his entire allowance on a bunch of new bacteria species trading cards. 2/3 of them were spirilla cards, 1/4 were bacilli, and the rest were rickettsia. If there were 20 fewer bacilli cards than spirilla cards, how many spirilla cards does Leonard have?

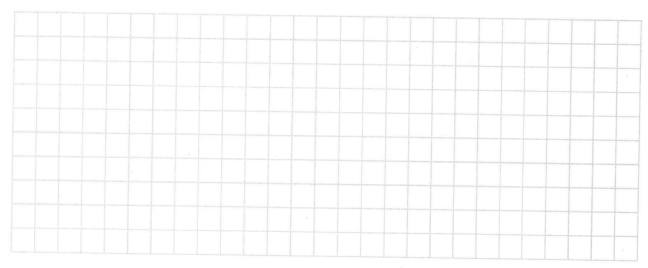

Leonard has _____ spirilla cards.

20. There are 360 students in Ghengis Khan High School's poetry club. The number of haiku fans is 3/4 the number of free-verse lovers. The number of free-verse lovers is 4/5 the number of students who really dig sonnets. How many free-verse fans are in the club?

There are _____ free-verse fans in the poetry club.

Name _____

21. Celebrity Dustin Dweeber ordered ⟨2⟩ pizzas from his favorite pizza shop for his posse. ⟨1-1/2⟩ of the pies will have a special gold-leaf topping and ⟨1/2⟩ pie will have plain cheese topping for Angus, Dustin's bodyguard, who doesn't eat precious metals. Dustin asked the shop to cut each pie into sixths. How many slices of gold-leaf pizza will the pizza shop make?

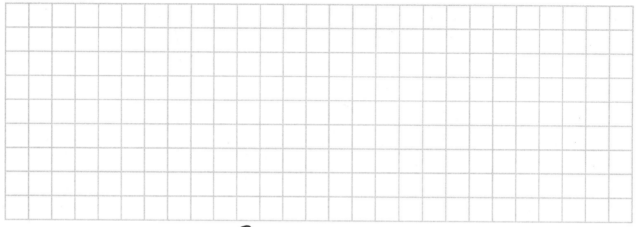

The pizza shop will make ___9___ slices of gold-leaf pizza.

---

22. Sports nutritionist Dr. Ferdie Fumble has developed a new bologna-flavored sports drink. His first test batch is ⟨3/4⟩ of a liter. He has determined an athlete should drink ⟨3/16⟩ of a liter before each game. How many 3/16-liter servings can he pour from his test batch?

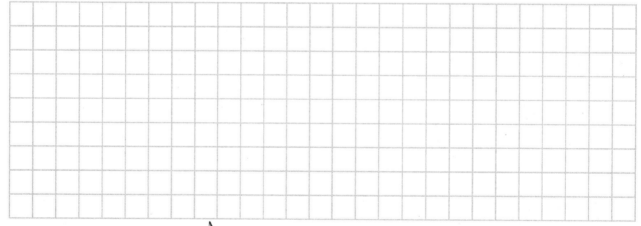

Dr. Fumble can pour ___1/4___ 3/16-liter servings from his test batch.

Name _____

23. Chef Uta made her famous banana pork loaf. The loaf is (2-1/2) feet long. If she cuts it into 3-inch slices, how many slices can she get from this loaf?

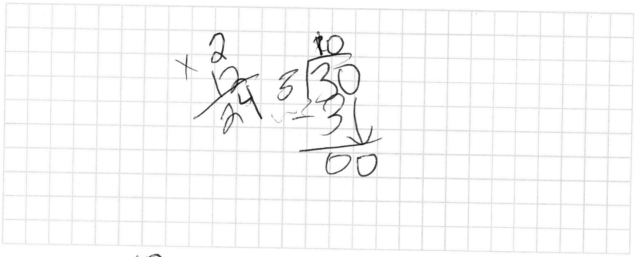

Uta can cut __10__ slices.

---

24. Mortimer is convinced that the newest sports craze will be snail racing. After an exhaustive breeding program, he has developed Lightning, a snail who can cover 8/8 of a foot in 10 minutes. How long will it take Lightning to complete a race on a (1-1/2) foot course?

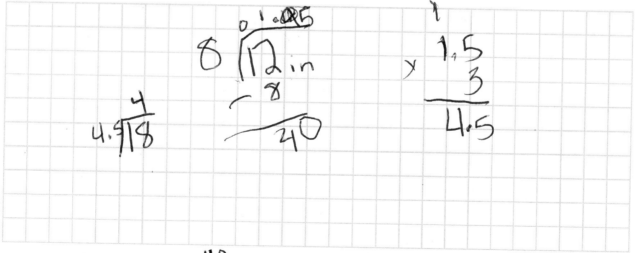

It will take Lightning __40__ minutes to complete the course.

Name _____

25. Max, Mabel, and Minky each have a collection of old Martian zloxny coins. Mabel has 2.5 times more coins than Max, and Minky has 1.5 times more coins than Mabel. If they have 638 zloxny coins in all, how many does Mabel have?

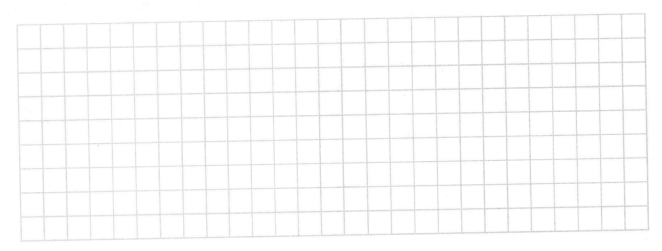

Mabel has _____ zloxny coins.

26. Quikster and Speedy work at Harris's Hotcakes 'n' Haggis food truck. Quikster earns $10.50 more than Speedy each week. They each spend $50 per week on yummy school lunches and donate the rest to the school's triangle and kazoo marching band. When Quikster's donations totaled $300, Speedy's totaled $90. How much does Speedy earn each week?

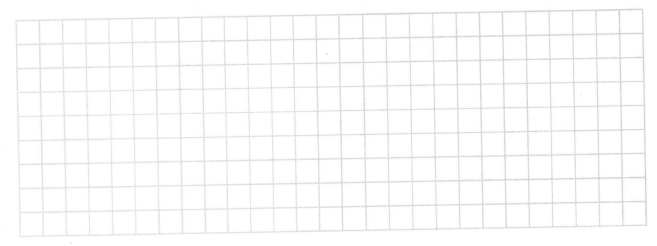

Speedy earns $ _____ each week.

Name _____

27. Terry the tarantula is 2.5 times longer than his sister, Teresa. If Teresa is 10.5 centimeters shorter than Terry, how long is Terry?

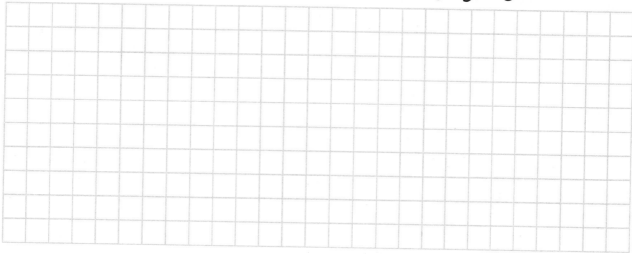

Terry is _____ centimeters long.

28. Rita and Ralph had to sell boxes of chocolate-covered asparagus to raise money for their triangle and kazoo marching band. Rita picked up 1.5 times as many boxes as Ralph. After Rita sold 70 boxes and Ralph sold 20, they had the same number of boxes left. How many boxes did Rita pick up to start?

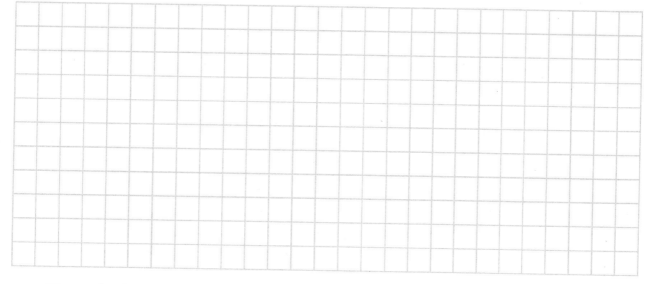

Rita picked up _____ boxes.

29. Tina Trashtruckian, celebrity extraordinaire, ate lunch at her favorite Beverly Hills restaurant, Le Petite Snob. She wasn't going to leave a tip, but her agent persuaded her to, so she left $184.56. That included the cost of her meal plus a 20% tip. What was the cost of the meal before the tip?

The cost of the meal before the tip was $ _____.

30. Alvin bought a bunch of fleas at the local flea market. He graciously gave 15% of them to his sister, Allie. If he gave 54 fleas to Allie, how many did he buy originally?

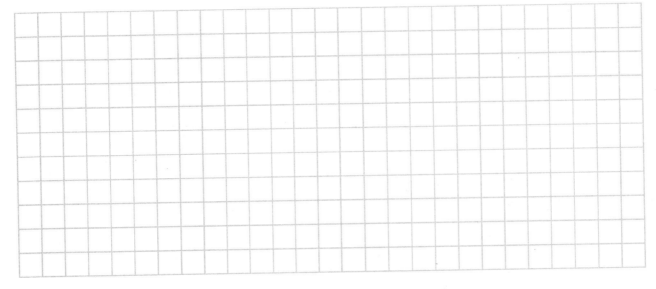

Alvin originally bought _____ fleas.

Name _____

31. Vesuvius got so excited when he heard that there was a 25% off sale on guacamole donuts at the Carb Shoppe, he almost blew his top. Instead, he paid only $1.50 for each donut. What was the cost of the donut before the sale?

The cost of the donut before the sale was $ _____.

32. Felix was a product tester at the Happy Fun Novelty Company. He had to test a batch of joy buzzers. 4% of the joy buzzers he tested—a total of 14—were defective. How many joy buzzers did he test?

Felix tested _____ joy buzzers.

33. Antiques aficionado Lulu recently appeared on *The Old Stuff Show*. The expert appraiser valued her vintage collection of disposable dishware, including 2 used Styrofoam cups, 2 mint-condition plastic forks, and an empty milk carton, at a total of $52,000. A fork's value is 50% as much as a cup, and the milk carton is valued at $3,000 more than a fork. What is the value of the milk carton?

The value of the milk carton is $ _____.

34. Red bought some supplies for a trip to visit her grandmother. She spent 0.4 of her money on a picnic basket filled with snacks and 5/12 of the remainder on wolf repellent. The wolf repellent cost $30.00. How much money did Red start with?

Red started with $ _____.

Name _____

35. 70% of the Varsity Hopscotch Team at Warren G. Harding Middle School are boys. If there are 60 more boys than girls, how many students are on the team?

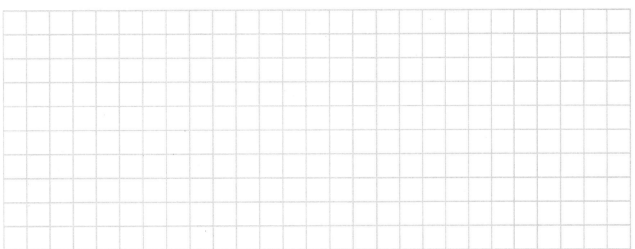

There are _____ students on the team.

....

36. Art Arachnid, lead singer with the Beetles, received an advance on their new release, *Don't Bug Me, Man*, from his record label. He spent 5/9 of his money on a princess-style horse-drawn carriage, gave 1/4 of what was left to the Society for Wayward Walruses, and still had $60,000 left over. How much was the advance from the label?

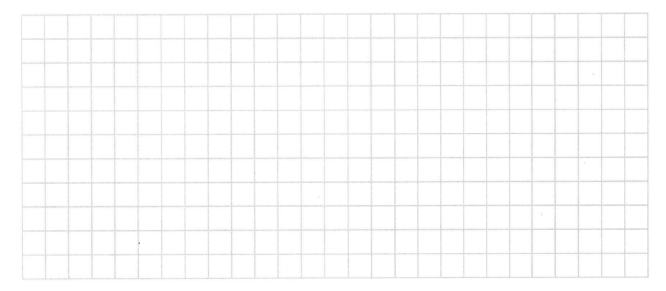

Art's advance was $ _____.

37. Renata's recipe for her famous Kale 'n' Clam Quencher calls for kale juice and clam juice in a 3:1 ratio. If Renata pours 18 liters of fresh kale juice into a vat, how many liters of clam juice should she add?

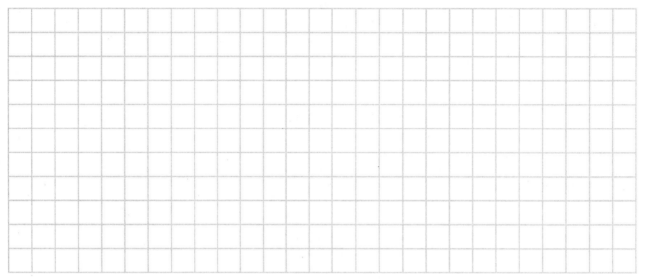

Renata should add _____ liters of clam juice.

38. Little Clara has 20 pets in all. Some are warthogs, and some are cobras. If the ratio of cobras to warthogs is 7:3, how many cobras does Little Clara have?

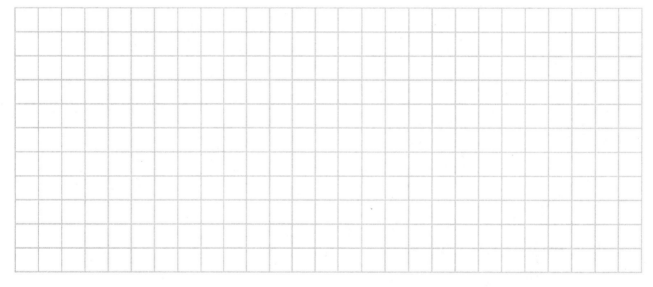

Little Clara has _____ cobras.

Name _____

39. Principal Jillian Genius runs the gifted program at the Wee Wise Ones preschool. The ratio of organic chemistry classes to Latin classes per month is 4 to 3. If she scheduled 12 organic chemistry classes in October, how many Latin classes were there?

There were _____ Latin classes.

40. Tina Trashtruckian is the celebrity famous for, well, being a celebrity! She recently ordered a larger-than-life painting of herself to display in her living room. If the painting has a perimeter of 400 inches and the ratio of length to width is 7:3, what is the painting's width?

The painting's width is _____ inches.

41. At the end of a hard day of interplanetary conquest, evil Galactic Emperor Zog relaxes with a tall glass of his special chocolate milk. He mixes 16 ounces of fine Belgian chocolate syrup with 4 ounces of whole milk. What is the ratio of chocolate syrup to milk? What fraction of the chocolate milk consists of milk?

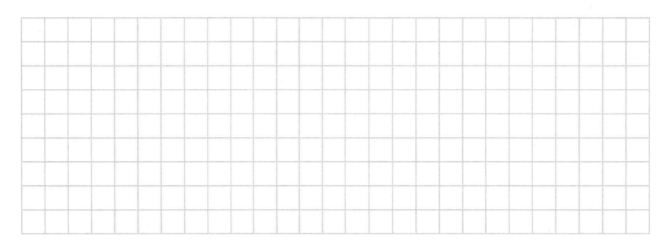

The ratio of chocolate syrup to milk is _____.

The fraction of the milk in the chocolate milk is _____.

42. Milton and Sydney are the two leading scorers on Slobovia's Olympic Jacks Team. Milton has scored half the team's points, and Sydney has scored 1/6 of the team's points. What is the ratio of Milton's points to Sydney's points?

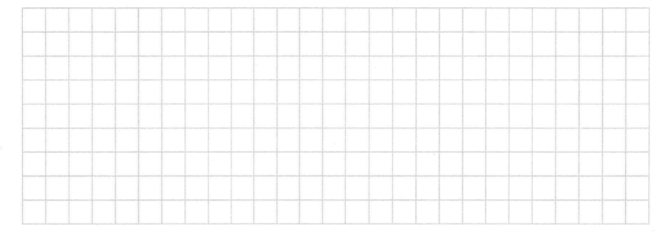

The ratio of Milton's points to Sydney's points is _____.

Name _____

43. Alberta is writing a report for social studies class. She doesn't have much information, so she figures she'll just add a lot of pictures. (Don't try this at home, kids!) If the ratio of text pages to picture pages is 1:5, what fraction of the report is text?

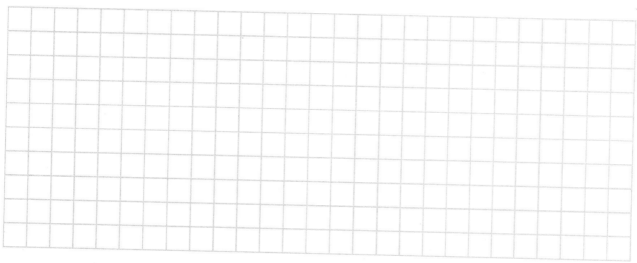

The fraction of the report that is text is _____.

44. Clarissa, the 12-year-old musical genius, has just written her latest masterpiece, "Symphony for Triangles and Bassoons." The score indicates that 2/7 of the orchestra should be triangles. What is the ratio of triangles to bassoons?

The ratio of triangles to bassoons is _____.

Name _____

45. Musicians Marvin, Millie, and Meriwether are raising money to fund the school's banjo ensemble's overseas tour. Altogether they have received donations totaling $600. If the ratio of money raised by Marvin, Millie, and Meriwether is 7:5:3, how much did Meriwether raise?

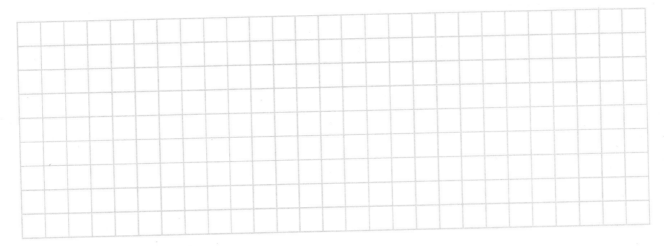

Meriwether raised $ _____.

46. Selene's Gourmet Frozen Yogurt shop features "specialty" flavors. Yesterday, she sold servings of Mustard Ripple, Pepperoni, and Bacon in a ratio of 2:5:4. If she sold 36 servings of Mustard Ripple, how many servings did Selene sell in all?

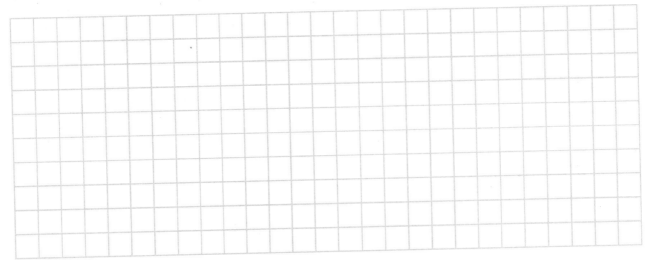

Selene sold _____ servings in all.

Name _____

47. Fashionista Farrah has the most awesome hat collection!
The ratio of the number of jesters' hats she owns to the
number of her Scottish golf hats is 3:1. The ratio of the
number of her Scottish golf hats to the number of her
rhinestone-studded biker caps is 3:4. What is the ratio
of jesters' hats to rhinestone-studded biker caps?

The ratio of jesters' hats to rhinestone-studded biker caps is _____.

....

48. Michaela's critter collection includes grubs, pigeons, and ferrets in a ratio of 4:5:6.
If she has 36 ferrets, how many grubs will she have to buy so that the number
of grubs equals the number of pigeons?

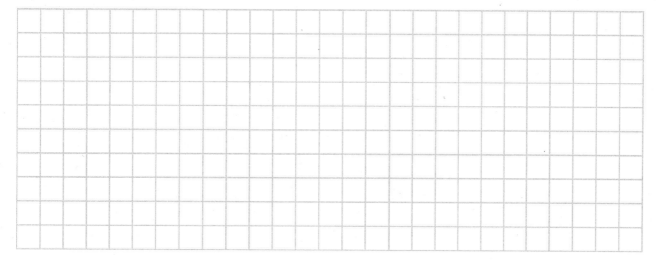

Michaela needs to buy _____ grubs.

Name _____

49. Grandma Gertie's special birthday-cake icing recipe calls for 4 ounces of maple syrup and 5 ounces of blue cheese. If she decided to make a larger batch of icing and used 20 ounces of blue cheese, how many ounces of maple syrup would she need?

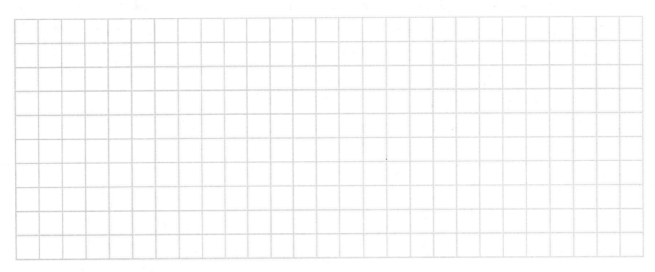

Grandma Gertie would need _____ ounces of maple syrup.

50. Lonesome Lenny Lewis, the town's most unpopular radio DJ, has decided to play only bagpipe tunes and accordion songs to appeal to the "young people." Each hour, he plays 6 bagpipe tunes and 4 accordion songs. Last night, he played 25 songs in all. How many featured the accordion?

_____ songs featured the accordion.

Name _____

51. A national survey found that teens prefer square
dances to minuets by a ratio of 3:4. If 279 teens
in Doughnut City prefer square dancing, and the
national averages hold true, how many teens prefer
to dance the minuet?

_____ teens prefer to dance the minuet.

52. Bettina's delicious trail mix includes chocolate drops, rhubarb jellybeans, and beef
jerky bits in a ratio of 3:4:8. If she uses 160 grams of rhubarb jellybeans, how many
grams of trail mix would she make?

Bettina will make _____ grams of trail mix.

53. Barry Sty, lead cowbell player in the band Pigpen, has a
large collection of industrial vehicles that includes tanker
trucks and cement mixers in a ratio of 4:5. He has 27
vehicles in all. If he decides to trade in 3 tanker trucks
and buy 3 more cement mixers, what will the new ratio
of tanker trucks to cement mixers be?

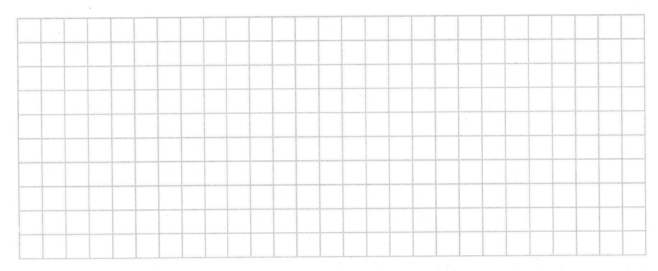

The new ratio of tanker trucks to cement mixers will be _____.

---

54. Rhonda and Miriam once had an equal number of pieces of pepper taffy.
Rhonda ate 10 pieces, and Miriam ate 6 pieces. Now the ratio of Rhonda's taffy
to Miriam's taffy is 2:3. How many pieces of pepper taffy did they each start with?

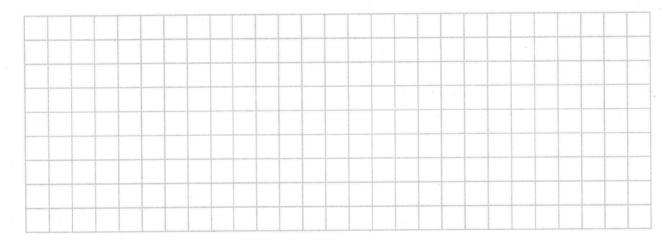

They each started with _____ pieces of pepper taffy.

Name _____

55. The ratio of hopscotch players to foursquare players at Ahab Elementary School is 2:3. If half the hopscotch players decide to leave the team and play foursquare, there will be 180 more foursquare players than hopscotch players. How many students are on the two teams?

_____ students are on the two teams.

56. Lamar and Lex each have an amazing armadillo collection. The ratio of the number of Lamar's critters to Lex's critters is 4:3. After 7 of Lamar's critters escaped (much to the chagrin of his parents!), they have an equal number of armadillos. How many armadillos did Lamar have at first?

Lamar had _____ armadillos at first.

57. Lester the forest gnome has a total of 20 pet toadstools, which is 7 more than what his sister Lexie has. How many pet toadstools does Lexie have?

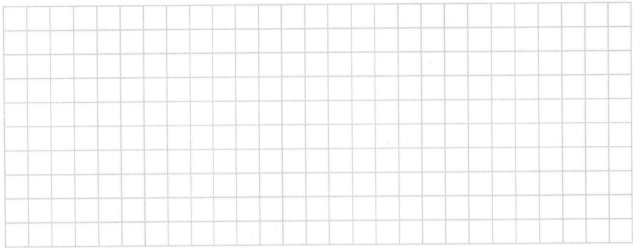

Lexie has _____ pet toadstools.

58. Maynard Mountain, quarterback for the New Jersey Noodles, owns 17 antique umbrellas. Nick Nicknack, the wide receiver, has 10 more than twice the number of umbrellas Maynard owns. How many umbrellas does Nick own?

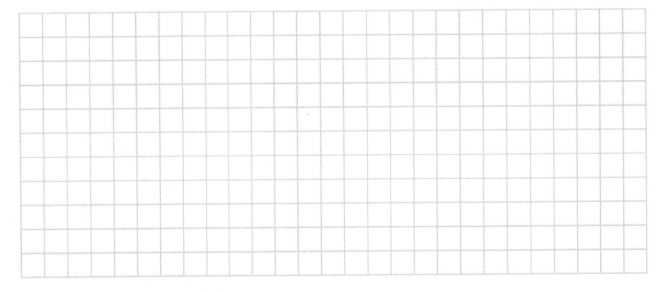

Nick owns _____ umbrellas.

Name _____

59. Silvia planted some artichoke thistles in her backyard. Each plant produces 35 pounds of artichokes. Her total crop weighed 420 pounds. How many artichoke thistles did Silvia plant?

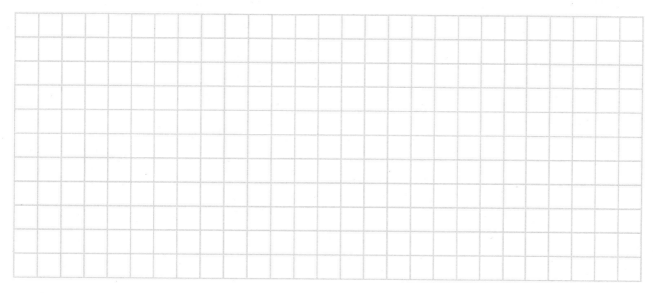

Silvia planted _____ artichoke thistles.

60. Princess Eloise started working at the fertilizer store. She gets paid $10 per day plus 1/5 of everything she sells. On her first day, she sold $250 worth of fertilizer. How much was Eloise paid on her first day?

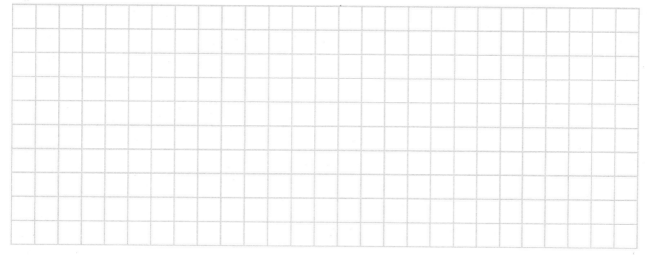

Eloise was paid $ _____.

Name _____

61. Arsene was transporting 5 containers of yummy turnip pie to be sold at his pie shop. Each container held 12 pies. While watching his favorite cartoon, *Super Gnat*, Arsene accidentally dropped a container, ruining 2/3 of the pies in the container. How many saleable pies were left?

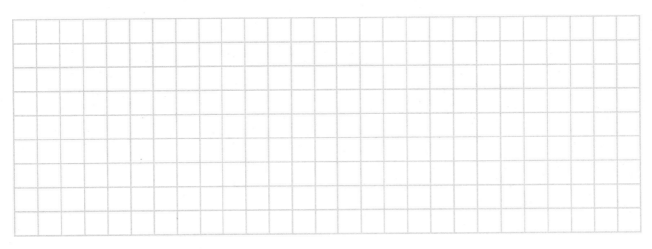

There were _____ saleable pies left.

62. Percy Pig was getting ready for a "visit" from B.B. Wolf. He ordered 4 truckloads of sticks to build his house out of flimsy, but easy-to-assemble sticks. His more practical brother, who had advised him to use brick, took pity on him and gave him 72 additional sticks. Percy now has 552 sticks. How many sticks were in each truckload?

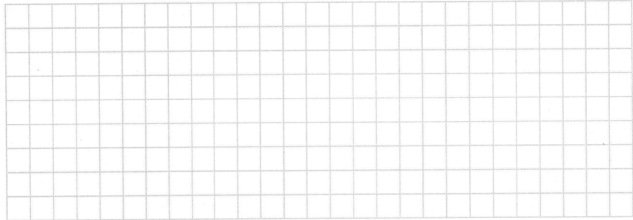

There were _____ sticks in each truckload.

Name _____

63. Igor bought 6 neck bolts and 1 extra, extra, extra large pair of basketball shorts. Each neck bolt cost $2.50. If Igor paid $32.50, how much did the basketball shorts cost?

The basketball shorts cost $ _____.

64. Tina Trashtruckian has decided to run for president on a "Selfies for All" platform. Upon her announcement, 70% of her fan club volunteered to work on the campaign. One month later, 20% of her volunteers quit. The next month, 250,000 new volunteers joined when she posted a "clever" comment on her Twit account. If the committee now has 330,000 volunteers, how many joined originally when she announced her candidacy?

_____ people originally volunteered for the campaign.

65. Lucretia went to a local craft fair to buy her mother birthday presents. She looked at a macaroni plaque, a garlic-scented bag of sachet, and a statue of a polar bear carved from a giant marshmallow. The 3 items cost $104 in all. The plaque cost $14 more than the sachet, and the sachet cost twice as much as the statue. In the end, she decided to buy 2 plaques and 2 statues. How much did she spend?

Lucretia spent $ _____.

66. There are 360 students in the Pretend Creatures Club. For every 5 members who love unicorns, there are 7 members who fancy trolls. How many more troll lovers are there than unicorn lovers?

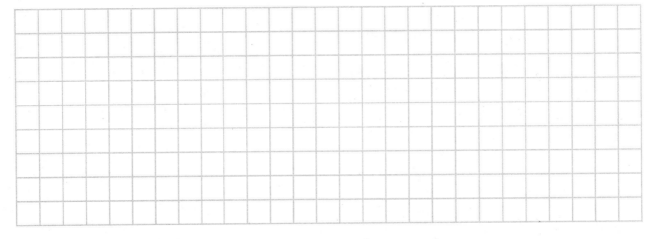

There are _____ more troll lovers than unicorn lovers.

Name _____

67. Dustin Dweeber brought samples of his new colognes to a talk show to give out to the audience. He had 240 bottles. 1/3 of them contained the fragrance Immaturity, 3/8 of the remainder contained Talentless, and the rest were his signature cologne Spoiled Rotten. What percentage of the bottles contained Talentless?

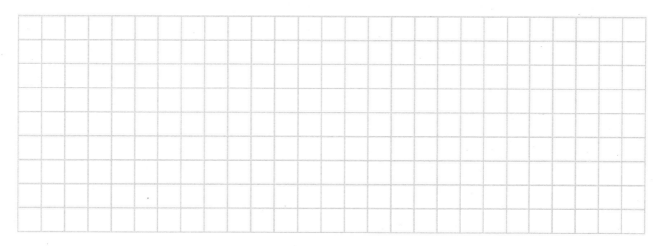

_____ percent of the bottles contained Talentless.

68. Bumbling magician Marvin the Marvelous thought it was time to present his act to the public. He set up a ticket booth outside the theater. On the day before the show, he sold 1/4 of the available tickets. On the day of the show, he sold 1/3 of the remaining tickets. He had 60 tickets left. What percentage of the available tickets did he sell?

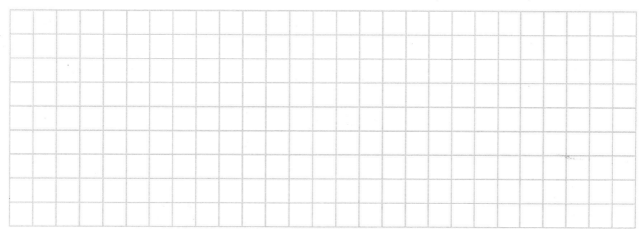

Marvin sold _____ percent of the available tickets.

Name _____

69. Bumbling magician Marvin the Marvelous was having trouble with his pull-the-badger-out-of-a-hat trick, so he decided to practice it 100 times per day. On Monday, he was successful 4 times; on Tuesday, 12 times; on Wednesday, 19 times; and on Thursday, 6 times. If his overall success rate was 10% for a 5-day period, how many times was Marvin successful on Friday?

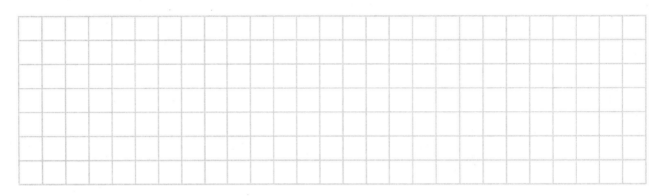

Marvin was successful _____ times on Friday.

----

70. The athletic director at Count Dracula Middle School ordered an equal number of uniforms for each of the school's 3 sports: swimming, gymnastics, and country-line dancing. There are 147 athletes in all. To get an equal number of athletes on each team, she moved 14 swimmers to gymnastics, 18 gymnasts to country-line dancing, and 12 dancers to swimming. How many athletes were on each team before the big switch?

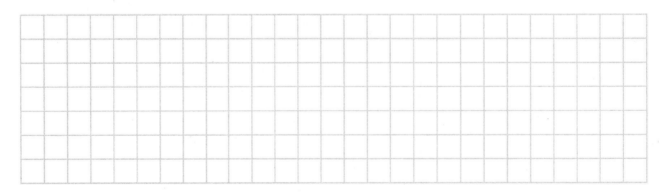

The starting number of swimmers was _____.

The starting number of gymnasts was _____.

The starting number of dancers was _____.

Name _____

71. Lothar spent the day at the Sword and Sorcery Shoppe.
He bought 4 cans of dragon repellent and 2 magic amulets
for a total of $37.40. If he had bought 8 cans of dragon
repellent and 6 amulets, he would have spent $89.60. How
much does each can of dragon repellent cost?

A can of dragon repellent costs _____.

---

72. Hadrian was working the night shift at Harris's Hotcakes 'n' Haggis food truck. He
was preparing the "secret sauce" for the next week. Hadrian made a batch of the
stuff and poured it into 7 large containers, 1 for each day of the coming week. On
Monday, the truck used 7.5 liters and had 1.25 liters left. On Tuesday, the truck
used 8.25 liters and had 0.5 liters left. How much secret sauce did Hadrian make?

Hadrian made _____ liters of secret sauce.

Name _____

73. Tina Trashtruckian started a new fashion trend—toting around miniature squirrels in a bag. For a total of $1,310, she bought 3 Nantucket red squirrels, 2 Princetonian black squirrels, and 1 regular old gray squirrel. A red squirrel cost 3 times as much as a black squirrel, and the gray squirrel cost $50 more than a black squirrel. How much did the gray squirrel cost?

The gray squirrel cost $ _____.

---

74. Rufus can't wait for lunchtime. Today is quiche day! The cafeteria made 3 times as many beet quiches as squid-ink quiches and 50 more kelp quiches as squid-ink quiches. They made 945 quiches in all. How many more beet quiches are there than kelp quiches?

There are _____ more beet quiches than kelp quiches.

Name _____

75. Quincy paid $750 for a limited edition Dustin Dweeber fanzine and 4 of his CDs. The fanzine cost 6 times as much as a CD. How much more does the fanzine cost than a CD?

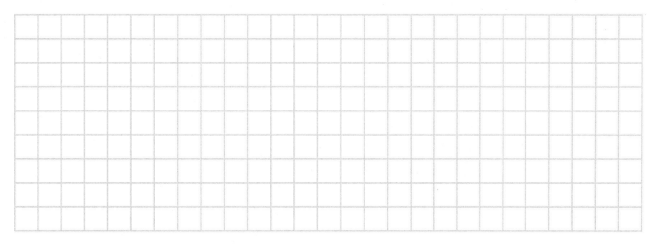

The fanzine costs $ _____ more than the CD.

76. Little Clara bought 2 pair of socks and 1 catcher's mask actually worn by Tina Trashtruckian! The catcher's mask cost 9 times as much as a pair of socks. Little Clara gave the store $2,200 and received $22 back in change. How much more does the catcher's mask cost than a pair of socks?

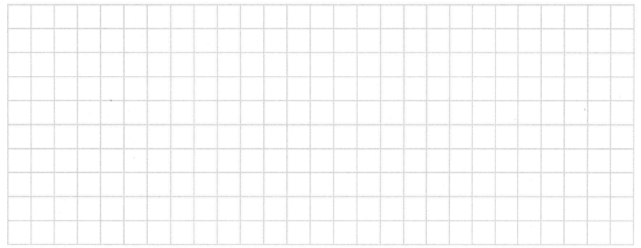

The catcher's mask cost $ _____ more than a pair of socks.

Name _____

77. Kyra was excited to win a place in this year's *R Games.* (*R* stands for Ridiculous!) In her medley event, she hopped on a pogo stick for 15 minutes at a rate of 8 km per hour and then rode her big-wheel tricycle for 1/2 hour at a rate of 6 km per hour. How far did she travel?

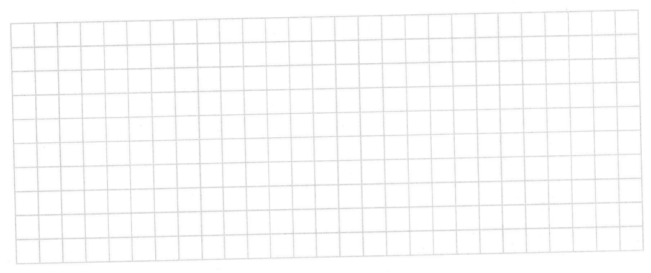

Kyra traveled _____ km.

78. Cassandra jet skied from Rotting Fish Beach to Polluted Point. She went 2/5 of the way in 2 hours. Then she shifted into high gear and went the rest of the way in 2 hours at 60 miles per hour. What was the average rate of speed for the whole trip?

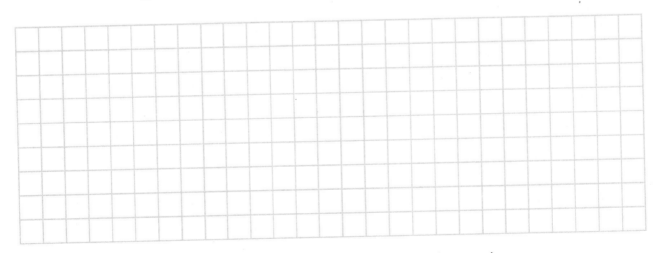

The average speed for the whole trip was _____ miles per hour.

Name _____

79. Calpurnia was training to improve her time for the crawl. (Not the swimming-type crawl, the baby-type, hands-and-knees method of propulsion.) On the first day of training, she completed 3 crawls of 100 meters each. The first practice crawl took 32 seconds, and the second crawl took 28 seconds. Overall, her average time was 29 seconds. How much time did it take her to complete her third practice crawl?

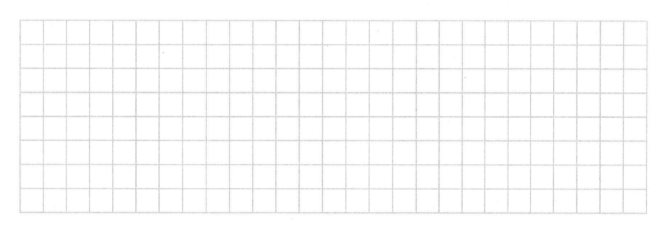

It took Calpurnia _____ seconds to complete her third practice crawl.

80. Celebrity Tina Trashtruckian was walking down the red carpet at the opening night of a new Hollywood blockbuster film. (Tina wasn't in the movie—she's a celebrity, not someone with acting ability!) When the photographers were snapping pictures, Tina walked the 360 feet from her limousine to the theater in 5 minutes. On the way back, when the photographers were gone, she walked to her limo in 1 minute. What was Tina's average speed?

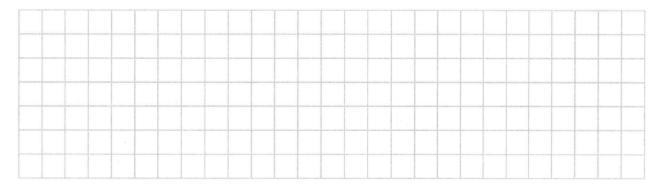

Tina's average speed was _____ feet per minute.

Answer Key

1. 1,444 pieces
2. 788 fezzes
3. 1,134 customers
4. 624 orders
5. 81 pictures
6. 13 beets
7. $189
8. 8 months
9. 3 wolverines
10. $80
11. 122 cheesesteak-flavored gumballs
12. 20 aluminum crowns
13. $70.70
14. 600 humorless comedians
15. $19.18
16. 9 books
17. 150 pieces of cheese
18. 72 centimeters
19. 32 spirilla cards
20. 120 free-verse fans
21. 9 slices
22. 4 servings
23. 10 slices
24. 40 minutes
25. 220 zloxny coins
26. $54.50
27. 17.5 centimeters
28. 150 boxes
29. $153.80
30. 360 fleas
31. $2.00
32. 350 joy buzzers
33. $10,000
34. $120
35. 150 students
36. $180,000
37. 6 liters
38. 14 cobras
39. 9 Latin classes
40. 60 inches wide
41. 4:1; 1/5
42. 3:1
43. 1/6
44. 2:5
45. $120
46. 198 servings
47. 9:4
48. 6 grubs
49. 16 ounces
50. 10 songs
51. 372 teens
52. 600 grams
53. 1:2
54. 18 pieces
55. 300 players
56. 28 armadillos
57. 13 pet toadstools
58. 44 umbrellas
59. 12 artichoke thistles
60. $60
61. 52 pies
62. 120 sticks
63. $17.50
64. 100,000 people
65. $136
66. 60 more troll lovers
67. 25%
68. 50%
69. 9 times
70. 51 swimmers, 53 gymnasts, and 43 dancers
71. $5.65
72. 61.25 liters
73. $155
74. 308 more beet quiches
75. $375
76. $1,584
77. 5 kilometers
78. 50 miles per hour
79. 27 seconds
80. 120 feet per minute